AF882469

SOTHIS PRESS

Sothis Press aims to completely reimagine the public domain book with an emphasis on universal access to captivatingly designed print editions. We believe that technology will enable every single book ever published to be perpetually accessible in beautiful print form. Sothis is beauty in ubiquity.

www.sothis.press
1 Waterhouse Square, London, ECIN 2ST

Coal and Coke

An Account of the Holiday Excursion of the Boys and Girls Among the Coal Mines

La; Man

SOTHIS PRESS

ISBN 978-1-0368-0807-5

This edition is published by
Sothis Press © 2023. The company
is registered in England and Wales.
The Company Number is 14900389.

AN ACCOUNT OF THE HOLIDAY EXCURSION OF THE BOYS AND GIRLS AMONG THE COAL MINES, BY

A MAN

OF THE GREAT ROCK ISLAND ROUTE.

Respectfully Dedicated to the Boys and Girls of America, by the Chicago, Rock Island & Pacific Railway.

CHICAGO:
THE J. M. W. JONES STATIONERY AND PRINTING COMPANY,
1888.

"Coal is entitled to be considered as the mainspring of our civilization. By the power developed in its combustion all the wheels of industry are kept in motion, commerce is carried with rapidity and certainty over all portions of the earth's surface, the useful metals are brought from the deep caves in which they have hidden themselves, and are purified and wrought to serve the purposes of man. By coal night is, in one sense, converted into day, winter into summer, and the life of man, measured by its fruits, greatly prolonged. Wealth, with all the comforts, the luxuries, and the triumphs it brings, is its gift. Though black, sooty, and often repulsive in its aspects, it is the embodiment of a power more potent than that attributed to the genii in Oriental tales. Its possession is therefore, the highest material boon that can be craved by a community or nation.

"Coal is also not without its poetry. It has been formed under the stimulus of the sunshine of long-past ages, and the light and power it holds are nothing else than such sunshine stored in this black casket to wait the coming and to serve the purposes of man."—PROF. J. S. NEWBERRY.

* * * "Above all, we should see another instance of the wisdom and goodness of Him who hath in so wonderful a manner placed beneath our feet inexhaustless quantities, and in an imperishable form the 'stored-up fuel of a world.'"—MCFARLANE.

INTRODUCTION.

With the return of the joyous holiday season, the GREAT ROCK ISLAND ROUTE adds another to its series of popular little volumes—this being the fourth. It is, in a sense, a companion book to that of last year (PETROLEUM AND NATURAL GAS), treating as it does, and in much the same manner, of another and an even more valuable one of earth's hidden treasures.

Under the title of "COAL AND COKE," the present number aims to present the most important facts concerning coal, its origin, its distribution, the methods of mining it, preparing it for and transporting it to market, etc., etc. To present such a theme, or set of themes, in style and language that shall be readily understood by the boys and girls, to whom the book is dedicated, has been no light task, and older readers will, it is hoped, pardon the frequent repetitions and the oft-times round-about methods of presenting matters, deemed necessary to make various points clear to the youthful mind. It has been the aim ever kept in mind to tell the story of coal in the most plain and simple, yet entertaining and instructive manner possible, all circumstances considered.

For very much of our national wealth and importance we are indebted to our wonderful coal deposits, and the same is true of England and her coal beds. The billowy blackness which rolls skyward day and night, from tens of thousands of great smoke-stacks, all over the land, attest unmistakably to the fact that our coal-beds lie at the foundation of very much, indeed, of our commercial greatness. This little book aims to afford a glimpse at the operations which transform our hidden, slumbering coal beds into one of the most important factors, if not *the* most important factor, entering into the material progress of the wonderful nineteenth century. It is hoped this glimpse of mining operations, including the scenes and incidents of a day spent "down in a coal mine," of the "towing" of coal on some of our great rivers, of the manufacture of coke, and of illuminating gas, etc., may be pleasing to every one who reads the book.

Again wishing you a "Merry Christmas and a Happy New Year," I remain

YOUR FRIEND,

"A MAN."

BOAT PASSING THROUGH LOCK.

I.

HOW THE EXCURSION WAS PLANNED.

ERE you all are, again? Well, well, I should say you are all here again, sure enough! Ned, Nell, Tom, Miss Inquisitive, and the rest; and not only so, but I see a good many new faces in the crowd—or mob, I might well call it. Come in, come in—all of you; glad to see the new-comers, as well as to see my friends of former years, again. I am glad, too, that you haven't forgotten me, but have remembered me and have cared to come to visit me. Do you—Oho! that's your game, is it? I understand it all, now—ha, ha! You old acquaintances have gathered together all the other young folks you know; have brought them with you here; and now, straightway, almost before you are in the house, the whole party besieges and beseeches me to be taken upon another excursion! A very shrewd trick, indeed, youngsters, and in the hands of such a band of schemers and conspirators, I fear I will have to surrender and agree to another sight-seeing tour. But where—dear me, what a racket! one would think you had just been promised the most handsome Christmas present to be found, to hear and see such rejoicing! There, now, that will do. I was just about—Miss Inquisitive, *can't* you settle down at all?—just about to ask where we should go; however, as you have got me into this trouble, I'll just let you get me out of it, at least, so far as the decision as to where we shall go, and what to see, is concerned. Can't, eh? Come, put your heads together, and see if you can't. Put on your thinking-caps, and while you enjoy its warmth, look steadily and earnestly into the fire, among the glowing coals—a plan much practiced and highly praised by poets, philosophers and others, as greatly aiding hard thinking.

While you all are thinking as hard as you can, I'll take a little nap, and you can wake me up when you decide the—well, Miss Inquisitive, what is it? You've hit upon the plan, have you? Oh! just want to ask a question; I might have known that, to be sure. I see you haven't given up your old occupation; a great girl for questions, you are—a regular query-box. Well, what do you want to know this time? My! what a question! What makes that lump of coal burn? I might say that the fire makes the coal burn, but then you'd be sure to ask if it wasn't the burning of the coal that makes the fire, and we would be back at the starting point, again. Suppose, though, we put your question this way: Why does the coal burn? and then we will have it in better shape for an answer. I might, now, tell you, after the manner of the big books which your big brothers and sisters study, that coal burns because it is composed of highly inflammable and combustible elements; but to put it in plain, everyday language, which boys and girls of your age can understand, I will answer that coal burns because it is composed, or made up, we will say, in part of certain gases—oxygen and hydrogen—which ignite, or take fire, as I heard Ned say, a while ago—take fire and burn very easily and quickly. These gases and the other substances or elements of which coal —but dear me! I mustn't begin to tell you more particularly, now, about how these gases, etc., come to be in the coal, as that would make a long story, carrying us back to the time of the formation of the coal-bed from which these pieces have been taken, and our talking about it would interfere with the hard thinking going on amongst the others, you know. Some other day I'll tell you all—well, I declare! every one of you listening to what we two have been saying, when, as we supposed, you were busily puzzling your young heads over the question, Where shall we go? 'Go on with the coal story?' Oh, no; better not begin it now, as we would likely be led on from one point to another, talking about how coal was formed or made; where it is found; how it is mined or taken from the ground, and—why, that's just the thing, Ned! All in favor of an excursion among the coal mines, say *aye*. Whew! what a shout! Those opposed, say *no*. Carried, unanimously, as the big folks say. To the coal mines we will go, then, as soon as we can get ready. Miss Inquisitive, we owe to you, after all, through your question, the settlement of the where-to-go question. You shall have especial consideration shown you, hereafter."

II.

"WHERE SHALL WE GO?"

"Now, having decided to visit the coal mines, we must further decide upon the particular coal region or field to be visited. Shall we cross the ocean and visit the mines of England or—? Eh? *you* won't cross the ocean,—nor you?—nor you?—nor you?—nor—. Well, we may consider it settled that we won't undertake a sea voyage in our search for knowledge. To what coal-producing region of our own country, then, shall we go? There are several—Oh, no, Tom, it doesn't all come from the same place; I was just about to say that there are a number of coal-producing States in the Union—some producing but a small quantity, others, millions of tons, yearly. Can any of you name some of the coal-producing States? Yes, that's right—Indiana; what others? Yes, Pennsylvania, Virginia, Tennessee. Any more? You must have forgotten the 'chief products,' as given for some of the States in your geographies. A number of others, some of them producing large quantities of coal, as Ohio and West Virginia, might be named. Well, to which shall we go? Ned says, Pennsylvania. I recollect, as perhaps some of the rest of you do, that he was very much interested in the matter of coal mining last year, owing to the frequent glimpses he had of mines and miners, while going about among the oil wells. I suspect that it is a long cherished desire to visit some of those mines and to talk, perhaps, with some of those grimy-faced coal-diggers, that leads him to suggest that we go to Pennsylvania. Ah, I thought so. Well, your choice is a good one, Ned; not just because it suits your especial desires, but because in the matter of coal, as in that of oil, Pennsylvania offers the best field for studying the work of producing, shipping, etc., in all its departments. We would find in Pennsylvania, should we go there, such coal fields as we could find nowhere else—great fields too, of quite different kinds of coal. We could, too, see the wonderful coke region with its thousands of great coke ovens, and—Oh, dear! too many questions to answer at once; nor would we have time to answer them singly. I see no way out of the matter but to accept Ned's suggestion, and agree to accompany him to Pennsylvania. Does anybody move that we visit the coal regions of Pennsylvania? Ah! yes, and the motion seconded by a dozen. All in favor of the motion, say *aye*;—whew! no need to call for the *noes* in this case. It is decided, then, that we make a holiday excursion among the Pennsylvania coal mines, and that we start, say next—well, as soon as Miss Ruffleton and some of the other girls can get ready to go; the boys could all start this minute, for that matter."

III.

OUR LITTLE WORLD.

"As it is, generally, an advantage to be posted, in advance, in regard to matters with which we may have to deal, it will not be amiss, since we are now together, to spend the remainder of the afternoon in considering some of the facts and theories in regard to coal—how and when it was formed or made, where it is found, etc.; things which seeing it mined, shipped and used would not tell us.

At the outset, it may not be out of place to take a look at the world we live in—or upon, rather—as a whole, before we enter upon the examination of the facts and theories which, as we shall proceed, will narrow our field of observation until we may not only find ourselves confined to some one particular coal region, but may, indeed, even be shut up, for a time, in some dark, damp and dismal coal-pit or mine—'where,' as the song says, 'no ray of sunshine ever can be found.'

In the first place, we must bear in mind that our earth—the world, as we call it—is but a small affair, in a sense. Ha, ha! Miss Inquisitive, you think 'it's a pretty big place,' eh? You find plenty of room for questions, anyway; don't you? Well, it would be 'a big place,' indeed, if we did not know of so many other and vastly larger worlds, as we are told of them by the astronomers—the men who are familiar with the facts in regard to the sun, moon and stars. In other words, our world is but one of countless thousands of worlds which we see all about us—the stars, we call them. I said, all about us; they are all about us, that is, all around us, but at wonderful distances from us—millions, billions, trillions of miles from us; distances which we cannot comprehend at all. The sun, from which we get our light and heat, is 1,245,000 times as large as our earth—'big place' as it appears to Miss Inquisitive, and to all of us for that matter; while all the stars we can see, together with almost numberless others which can be seen only by the use of great telescopes, so great is their distance from us—all these, the astronomers tell us, are other suns, which may be as large as, or perhaps larger than, our own large sun. So great is the distance of some of these stars—or suns, as we are to think of them—that (so the astronomers tell us) the light which comes to us from them has been hundreds, if not thousands, of years in reaching us! although light travels, it has been learned, at the wonderful speed of 183,000 miles per second—or more than seven times the distance around our earth while you wink! You will understand these things of which I have just spoken, as well as learn very many more wonderful facts about our earth, the sun, moon, planets and stars, when you take up the study of astronomy; at present, however, we can not stop to talk about them, as we have other business on hand, you know. But the statements I have made are sufficient, in themselves, to show you that our world is but one—and a comparatively small one at that—of untold numbers of worlds which throng the space around us—the universe, we call it; it is as a single grain of sand in that great heap of sand and pebbles, which you see across the way!

But we started out to talk about coal; and why, then, should we go away off among the stars? you may be asking yourselves. Well, for several reasons. One of them is, the desire to set some of you to thinking, reading and studying about the wonders of astronomy—which I trust it may. Another is, to present a truth which I would have you keep in mind when we come to consider some facts in regard to our own earth, in particular,—that everything in the great universe, of which our earth is but a small part, as we have seen, is controlled or governed by certain laws ; that nothing is of chance or mere happening, but has been made what it is, as it is, and placed where it is, in accordance with a great plan or order of things, as designed and only fully understood by the great Creator.

As young astronomers, you will learn that not only is the universe crowded, as I have already stated, with almost countless worlds, but that all these worlds, as well as our own, are whirling or shooting through space at a most astonishing speed, and never stopping or pausing, either, in their wonderful flight! But you will learn more than this. You will learn that all their movements are controlled by certain laws; that they are not flying about helter-skelter, as it were, without direction or aim or purpose, and at the risk of frightful collisions with each other, but that all their courses are fixed, each having its own path or orbit, from which it can not stray or wander, and that throughout all the ages they have been flying, at most terrific speed, (our earth, more than 1,000 miles a minute!) around and around their great circular orbits, * and yet with a regularity which never varies even the smallest fraction of a second, so that thousands of years in advance astronomers can say where a planet or star will be, at a certain second! Wonderful, isn't it?

Now, there is a point in this, for us, taking it in connection with the littleness of our earth as compared with the vast universe. How the wonders of our own little world shrink and fade, as our thought carries us out among all the other and greater worlds! It is easy to see, now, how the One who could create all these other worlds, (tossing them out, as it were, into space, putting each in its place and appointing each its course through the heavens, and holding them all there by unseen power,) could create our one little world, just as it pleased Him—putting an ocean here, and a continent or an island there; hollowing out the valleys, building the mountains and spreading out the plains; clothing the earth with grass and flowers and trees, and filling it with beasts and birds; and finally peopling it, when at last made ready for man's home. As throughout all the rest of the great universe, so throughout our little world—a plan and purpose is everywhere seen. Our earth has been fitted up for man's use, and all that he needs has been prepared for him —prepared for him before he was created.

Among the things thus prepared for us, and of which we find an abundant store, is coal—to a consideration of some of the facts and theories concerning which we will now more directly give our attention."

* The orbits are not exactly circular, but are slightly *flattened circles*, we may be permitted to call them—ellipses, correctly speaking.

IV.

A WORD ABOUT GEOLOGY AND GEOLOGISTS.

"First, then, let us see what may be learned about how coal was formed. But before we begin, please punch that big lump of coal with the poker, Ned, and let's have a rousing fire. Now, come up closer, all of you ; there, that's both comfortable and cheerful, isn't it? Now, to begin.

You all know, I suppose—anyway, I've been taking it for granted that you all know—that coal is taken from the ground. It is not something made by men, but was prepared for men by GOD, in His wisdom and goodness, long ago—just how long ago we cannot tell. Ha ! ha ! at it already, Miss Inquisitive. How was it prepared, eh? Well, we must go to the geologists for an answer to your question, ma'am, as we did last winter when we wanted to know how oil and natural gas were formed or made. For the benefit of the new members of our company, I will repeat the explanation then made in regard to geologists and geology, adding, perhaps, a little more than then given. The word geology has been formed by the union of two Greek words—*ge*, meaning the earth, and *logos*, meaning a discourse, or talk, we may say ; the word geology, then, means a talk about the earth—its structure, etc., and of the changes which it has undergone. Geology is the science which treats of the earth's structure and its development or progress in the formation of rocks, progress in forms of life, animal and vegetable, etc. Geologists are the persons who have learned, and who give to us, the facts of geology.

What, then, can we learn from the geologists? Well, while they cannot tell us everything we may want to know, they can, at least, give us very much information, and some of it most wonderful, too, in regard to the formation, long ago, of the coal which is now being taken from the hills and the depths of the earth, and some of which is blazing and glowing, so much to our comfort, this chilly day, in the fireplace before us. And while we are thus enjoying its cheerful presence, I will try to present to you some geological facts, as made known to us by the learned geologists.

Those of you who were in our company last winter are already familiar with many of these facts, but you will not, I am sure, be unwilling to have them again stated, for the benefit of our new members, especially as I may speak more fully in regard to some of them, some having been, indeed, little more than merely mentioned before.

The geologists, then, tell us of great and wonderful changes which have taken place upon and within our earth. Even now, indeed, changes are occurring, by reason of earthquakes and other mighty forces; but the changes to which I refer took place long, long ago, before God had created people to live upon the earth—because it was not yet ready for them—and these changes were far more wonderful and powerful in effect than the earthquakes and volcanoes of the present time. Just how long ago these changes began which have resulted in producing the earth as we find it now, even the geologists cannot tell us; it is known only to Him of whom the great Psalmist wrote: 'Before the mountains were brought forth or ever thou hadst formed the earth or the world; even from everlasting to everlasting, thou art God.' The Bible tells us that at the beginning the 'earth was without form and void,' not the beautiful world it now is, with its hills and valleys, mountains and plains, its fields and forests, its rivers, lakes and oceans, and all the forms of life we see about us; all these things were then but on the mind of God, who was creating the world and fitting and preparing it for us, with all our varied wants and needs, when, after all else should be made ready, man should be created.

To learn something of the wondrous plan of the earth's creation and of its gradual perfecting, has been the study of geologists. Many facts have been established, while many theories or opinions have been offered, to account for some of these facts not fully understood. The geologists are able to tell us very much about the structure of the earth, as by long and patient search and study they have learned many things; but in regard to other things, they are in doubt and are not agreed among themselves. They can tell us what and how some things *are*, but differ with each other, and cannot speak positively as to how these things have come to be what they are, how they are and where they are. You will recollect, some of you, that this uncertainty was met with, a year ago, in our efforts to learn how petroleum and natural gas have been formed in the depths of the earth; and we may find the same uncertainty again, as we seek to learn all we may about the formation of coal. But while there are doubts and differences as to some points, we shall find geologists nearly all agreed touching what are given to us as the main facts in regard to the formation of coal.

While many things are hidden from the knowledge of man, so that he is not permitted to understand fully the great work and mystery of creation, yet it has pleased the great Creator to leave such signs and evidences of how very much of the great work which has given us the earth as it now is has been carried on, that the geologists, reading these signs and interpreting them, have given us various theories or opinions in regard to some of the great facts concerning the earth's formation,—some of which theories have come to be regarded as true and correct.

That our earth has not always been as it is now, but has passed through many very great and wonderful changes since first formed, is a fact agreed upon by all who have given the matter sufficient thought to have formed an opinion at all. These changes, however many and wonderful, have not been the result of chance or accident—that is, without direction and without aim or purpose; the contrary is the case. As I said, a little while ago, everywhere in nature is to be seen a wonderful plan, into which all of the great changes alluded to fit, each in its place, and all working toward one great end. As the learned Prof. Dana, an eminent geologist, says, an 'Infinite Mind has guided all events toward the great end—a world for mind,' and, 'the earth has, under this guidance and appointed law, passed through a regular course of history or growth.' Some of these changes are of especial interest to us in searching for the origin of coal and the manner of its formation, as now seen in 'beds;' and at these changes, with some others, perhaps, as the geologists describe them, we will glance.

Stir up the fire again, Nell— that's good! Now for the geology of the coal-fields."

V.

THE CARBONIFEROUS ERA--FORMATION OF THE COAL BEDS.

"That the hard, black lumps we see burning in the fire before us and which we call coal, is of vegetable origin--that is, that it is composed of the greatly changed remains of plants which grew and flourished long ago—is now perhaps a generally believed opinion. I say *perhaps*, because there have been some, and there may yet be some, who believe that coal has not been formed from old time plants, etc., but that it has been produced or made in another way; but there are so many evidences of its vegetable or plant origin, that few persons, if any, now believe any other theory as to its formation, or production.

But then, again, while geologists may agree as to the vegetable origin of coal, there are yet differences of opinion as to how the changes were brought about which turned—converted, the books would term it—the plants of that long-gone time into the 'beds' of coal from which we now so largely draw the supplies of fuel which warms our houses (and lights many of them, too,) and cooks our food, and which makes the steam that moves the machinery in nearly all of our thousands upon thousands of mills and factories, and sends the railroad trains speeding across the continents, and the steamboats and steamships up and down our rivers and over the pathless oceans. As concerning the formation of oil and natural gas—as we found last winter—so in regard to the formation of the coal beds,—there are different theories held by geologists, based upon what is, to them, satisfactory evidence, in each case. We will examine some of these theories, as briefly as the case will permit, for you must know that considerable geological history is to be touched upon in the course of such an examination.

To begin, then, I will state that the geologists divide the history of the earth's development—that is, the formation of rock beds, etc.—into a number of *ages*, the ages being again divided into *periods*, and these again, into *epochs*—all of which you will understand better when you are older. I may have to speak of these divisions in a somewhat general way, but need not stop to explain, particularly, in regard to them. Of the longest divisions, or *ages*, there are seven: First, the *Archæan*; second, the *Silurian;* third, the *Devonian;* fourth, the *Carboniferous;* fifth, the *Reptilian;* sixth, the *Mammalian;* seventh, the *Age of Man.* Each of these ages has its beds or *strata* of rock, very many of them, and of various thicknesses, several *strata* of the same kind occurring, with others between. The same kind of rocks are found, too, in different ages, and all these *strata* of all these seven divisions, taken together—a thickness of 20 miles, or more—make up the earth, so far as we see it and know it. The Archæan rocks, granite, gneiss, marble, etc., are the lowest rocks, forming the ocean bed, I may state. On these are the rocks of the next age, and so on, to the last rocks formed, at the top of all.

It is with the fourth, or *Carboniferous* age, that we shall have especially to deal in our study of the origin and the manner of formation of the coal-beds. Indeed, it is the presence of the coal-beds among the other rocks that gives the name *carboniferous* to this age; coal is composed largely of carbon, as we shall see, and hence the name *carboniferous*—abounding in carbon.

As just stated, the beds of coal are among—that is, they lie between—other rock *strata*, and—well, Mollie, what is it? Ah, yes, I've forgotten that some of you were not with us last year and, therefore, do not understand some of the statements I have made, or even some of the words used. Ned and the others of last year's party must wait here, now, until the rest of you catch up, as it were; but lest they grow impatient, I will make the necessary explanations as brief as possible.

All the beds of rock—hundreds of them, and of various kinds and thicknesses—which lie upon or above the lowest rocks—the granite, etc.,—are what are known as *sedimentary* rocks; that is, they were deposited where now found, by water, as sediment—sand, gravel, mud, etc. The geologists tell us that all, or nearly all, of what is now our dry land, was once—indeed, may have been many times—covered by water; the ocean now sweeping and spreading in, and again flowing off and out, as the land sank or rose by reason of earthquakes or other great forces. As the waters swept over the land, they brought with them sand, gravel, etc., broken or worn from the older rocks, and these, of course, settled to the bottom of the water, covering the rocks below. Sometimes, a great depth or thickness of this sediment would be thus deposited; while at other times a thinner bed of it would be formed. Thus, one bed upon another was deposited; and these, hardening under the pressure of those above them, and of the water itself, became beds of stone—mainly the beds of what we now call *sandstones*—(of various kinds, in the matters of color, hardness, etc.,) *conglomerates*, etc. The beds of limestone, which many of you, I'm sure, have seen, were formed in another manner—from the deposits of great quantities of shells of very tiny animals, with which the waters abounded; also, perhaps, from the remains of other small sea animals, such as the beautiful *crinoids*, or 'sea lilies,' as they are sometimes called, from their very great resemblance to lilies.

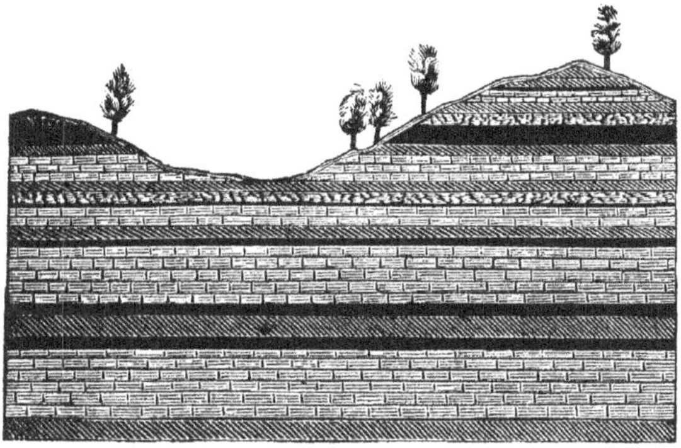

ROCK STRATA, AS IN BITUMINOUS COAL REGION.

Those of you who have had an opportunity to see the edges or faces of the rock-beds, in railroad cuts, quarries or steep hillsides, have noticed that the beds are not all of the same thickness, some being but shallow or thin, while others are of great thickness or depth; also, that the same bed may be thicker in some places than in others. These differences were caused by the difference in the quantity of sediment deposited at each time and place—in some cases but a comparatively small quantity; in others, a very large quantity. You will have noticed, too, that any bed is not one solid piece throughout its whole depth, but that it is made up of a number of layers, of various thicknesses, as we see in what we call 'flagstones,' which are taken, layer after layer, from the bed. Each bed, as made up of these layers, few or many, is called a *stratum;* when we speak of more than one bed we say, not *stratums*, but *strata*—the word which Mollie did not understand. Because of this arrangement of the rocks in regular beds or *strata*, they have been given the general name of *stratafied* rocks. Now, I think you all understand these points, and we will take up our story again.

Where were we? Oh, yes; I was just saying, when Mollie put her question, that we find the coal-beds, or *strata*, occurring among other *strata*—of sandstone, limestone, slate, shale, clay, iron-ore, etc. And the question which I see in the eyes of each of you is, How did they get there? How, indeed?

I have already stated, or have given you to understand, at least, that our coal-beds are found among the *strata* of what the geologists call the carboniferous age or era. Now, for a description of that period. The carboniferous era is divided by geologists into three periods—first, the *Sub-carboniferous;* second, the *Carboniferous* proper—or *coal-measures;* and, third, the *Permian.* The sub-carboniferous period (*sub* meaning under) was what may be called a time of preparation for what was to follow. The sea covered the land, and there were deposits of limestone, sandstones and shales, with a few thin seams of coal—which are sometimes called the *false coal-measures.* Then followed the true carboniferous period—the time of the formation of the great coal-beds, as we shall see. Following this, again, and closing this age or era, was the Permian period.

Passing over the sub-carboniferous period, which has been briefly described, let us enter the second or true carboniferous period, and see the formation of the coal-beds. At the beginning of this period we find the land covered with water, in which flourished myriads of crinoids or sea-lilies—which became the upper limestone of the sub-carboniferous period. Currents and waves now swept over and buried all under a great depth of sand and pebbles—which later hardened into the stone now known as the *millstone grit,* and which underlies the great coal-beds. This millstone grit is as much as 1,500 feet thick in parts of Pennsylvania. Above this, lying between various rock *strata,* we find the beds of coal—making up altogether, however, but about one-fiftieth ($\frac{1}{50}$th) of the total thickness of all the strata of the carboniferous, or coal-measure, period.

During this period there were many successive risings and sinkings of the land—the water sometimes being quite deep, and at other times barely covering the surface, with parts, here and there, not covered at all. It was during such times as those last mentioned—periods of shallow water, producing vast swamps or marshes—that the coal-beds, or what are now the coal-beds, were deposited. Over these marshes vegetation sprang up—a vegetation such as does not now exist, excepting to some extent, in the hot tropical countries. The temperature was much higher then than now, both water and air being very warm, the air being also very moist and highly charged—filled, that is—with carbonic acid, a gas composed of carbon and oxygen. These conditions of water and air produced a most wonderful growth of every form of vegetation then existing. You all are aware that trees, plants and all forms of vegetation grow most rapidly and luxuriantly during a season of especial warmth and moisture, but you are not aware, I suppose, just how they live, so I will tell you. They live mainly by means of carbonic acid, the gas of which I have just spoken, gathering it from the air through their leaves, which are their lungs. You can see, then, that the time of which I have been speaking, when the air was filled with carbonic acid, (the air now contains but a very small quantity of it) would be one especially calculated to promote the growth of all kinds of vegetation. And so, in truth it was. The great marshes of which I have spoken were dense jungles of plants and trees of the kind then growing. The trees were *lepidodendrids*, tall, and resembling in foliage, our pines; *sigillarids*, much like the first mentioned; some species of *conifers*, very soft pines, *tree-ferns*, etc., etc., while everywhere there was a dense growth of ferns of the low varieties, and of other vegetation. Growth upon growth of such vegetation occurred, until in some cases a great depth of old and decaying vegetable material would be formed. Then came a change, the land, with its profuse vegetation, being sunk beneath the water. Over the sunken forests or jungles were deposited beds of sediment—sand, gravel, etc., and of the small limestone-making shells, etc. Again, the land would be brought to the surface, and soon another marvelous growth of vegetation, as before, would occur—only to be in turn sunken and covered by beds of sediment, etc., until very many such beds of vegetation had been buried, some of them of great thickness or depth; others, of but little depth.

It is these beds of vegetation, thus long ago buried in the earth by GOD, in His great wisdom and goodness, that we now know as the coal beds. After being buried, as we have seen, beneath beds of sediment, the great weight above them, coupled with great heat, so acted upon these vegetable masses as to change them from the delicate and beautiful ferns and other plants into the solid black beds of coal—the beds of sediment, etc., deposited with them, becoming, at the same time, the beds or *strata* of sandstone, limestone, slate, etc., lying between and above the beds of coal. Thus our coal-beds were formed, according to what is probably the most generally accepted theory.

Other theories differ from this in some particulars. What is known as the *drift theory* is, that the vegetation which furnished the coal-beds instead of having grown and been buried, as described, where the beds now are, was carried to these places as drift material by waves and currents, from higher portions of the earth's surface, and then covered with the beds of rock material. Other theories present yet different views as to how the beds of vegetation were deposited and buried; but we will not stop to consider each, the one given being, as I said, a generally accepted one.

But however they were deposited, there can be no doubt in regard to the generally accepted belief in the vegetable origin of coal; the proofs of its correctness, as we shall see, are too conclusive to admit of doubt. Once great masses of vegetation, they have been changed into beds of coal, composed in great part of the carbon as held in the plants, combined with oxygen and hydrogen—the gases which cause the ready burning of the coal. Anthracite coal is almost pure carbon—more than nine-tenths carbon indeed. It may be interesting to you to learn too, that *graphite* (called also *plumbago* and '*black lead*'—though not lead, at all) and which is, familiar to us all as stove-polish, and as used in our lead pencils, is carbon; while the diamond, that most beautiful and valuable of gems, is after all, nothing but a bit of pure carbon—a very near relation, you see,

2

to the lump of coal which we burn up, or carelessly throw away. You can understand, then, the appropriateness of the name often applied to coal—'black diamonds!'"

VI.

PROOFS OF VEGETABLE ORIGIN — THE WORLD'S COAL FIELDS.

"That coal is of vegetable or plant origin, as we have seen, there is abundant proof—the main points of difference between geologists being as to how the beds were deposited and buried. We will let the geologists fight this matter among themselves, while we talk about the points concerning which they are agreed. Proofs of the vegetable origin of coal, then, are found on every hand. In the coal itself—some kinds, especially—can be seen the impressions of plants, ferns, etc., the delicate and beautiful leaves appearing as perfectly in this way as photographs, almost. Then, again, the trunks or stumps of trees, some of great size, are often found in the coal-beds. There is always immediately under the coal beds a *stratum* or bed of clay—called the *underclay*—which is full of the roots of the old plants; being, no doubt, the bed upon which the dense growth of vegetation took place, in the marshes spoken of. This clay, by the way, is used in the manufacture of fire-bricks—and is, therefore, often called fire-clay. The slate, shale, or other such material which immediately overlies the coal, and which when the coal is taken out, forms the roof of the mine—Never mind, Miss Inquisitive; you'll learn all about mines and roofs, later,—this slate, I say, is often covered with the prints of ferns, leaves, etc.,—all these things which I have mentioned pointing conclusively to the vegetable origin of our coal.

Some of you, no doubt, have noticed a difference in the kinds of coal you have used—Ah, yes; I thought so. Well, there are two principal kinds of coal—*anthracite* and *bituminous.* Anthracite is very hard; bituminous coal is soft—that is, as compared with anthracite. Between these two general divisions of coal, however, there are a great many kinds or grades of each. From the very hardest of coal, as found in some places, anthracite becomes less hard, in other places, until we have *semi-anthracite*—*semi* meaning half. So, also, with bituminous coals—they run through a great many grades of hardness—or softness—and of other qualities. All kinds or grades had the same origin, and were, no doubt, once all alike, in general, bituminous,—anthracite being but a changed and hardened form of bituminous coal, resulting from causes which we shall learn, later.

Coal everywhere? No, Ned, it is not found everywhere, though probably once existing in many sections where not now found. At the time of the deposits of the beds of vegetation, the surface of the earth was much more smooth and level, in general, than it is now, and the marsh-jungles covered vast regions. The wearing away and carrying to other places, as sand or soil, of the land—including rock and coal *strata*—by the rains and streams, has left some sections now barren of coal, though once it was present; so, we often find coal in the sides of the hills, while the low ground around has none, it having been carried away.

While speaking of this matter, I may as well tell you where the coal-fields are, in what countries found, their extent, etc. First in the list is our own country, containing the most extensive and valuable coal-beds in the world—covering an area of perhaps 150,000 square miles.* This extent of coal-land is not all in one great tract, but is divided up into a number of coal-fields or coal-regions as they are called. The greatest of these, as well as the greatest in the world, is the *Appalachian* field. It embraces the Pennsylvania region—which we propose to visit—covering a large part of the State, in anthracite and bituminous fields, and from thence, following the Appalachian system of mountains, (the Alleghenies, Blue Ridge, etc.,) it extends southwest over parts of Ohio, Maryland, West Virginia, Virginia, Eastern Kentucky and Tennessee, and across the northwest corner of Georgia, to the middle of Alabama, having an area of at least 60,000 square miles. The *Central* coal-field embraces a large part of Illinois, Southwestern Indiana and Western Kentucky, having an area of 47,000 square miles. The *Western* field embraces parts of Iowa, Nebraska, Missouri, Eastern Kansas, Arkansas, and extending into Texas, with an area of 78,000 square miles. Then there is the Michigan field—covering an area of 6,700 square miles, in the central part of the State. There is a small field in Rhode Island, extending slightly into Massachusetts, with an area of but 500 square miles. These are the principal coal-fields of our country, so far developed; but immense deposits of coal are being found throughout the far Western States and Territories, though of not so good quality as that of the eastern fields.

In the British provinces of Nova Scotia and New Brunswick, adjoining the United States on the northeast, there are valuable coal-fields, covering an area of 10,000 square miles. Central and South America, also, have coal fields, though small ones.

On the other side of the ocean, too, they have coal. England has long been noted for its production of coal—that is, the quantity mined, in this respect outranking even our own country, although we have more coal, having the greatest and best beds. Wales, also, is a great coal-mining country; while Scotland and Ireland also have their fields. The whole extent of the field of Great Britain—England, Wales, Scotland and Ireland—is but 12,000 square miles. Coal is also found in Spain, Portu-

* This does not include the later formations, or Tertiary coals, of the far Western States and Territories, and which are of very great extent. The total of all kinds for the United States is about 625,000 square miles.

gal, France, Germany, Austria, Russia and perhaps, to a lesser extent, other countries in Europe. Far off Australia, India and China, too, have coal-fields. So, you see, we do not have *all* the coal over here.

The greater part of all the coal of the world is bituminous, the soft coal. The fields of anthracite are few and small as compared with bituminous fields. The total area of all the anthracite fields of the world is said to be but about 2,000 square miles, and of this area nearly one-half is in the United States, Central Pennsylyania containing the greatest anthracite region of the world. The small coal-field of Rhode Island is also anthracite. Abroad, anthracite is found in Wales, Ireland and France.

Besides anthracite and bituminous coals—the true coals—there are beds of other kinds of coals, or substances used as coals. *Brown coal* is a brownish-black coal of the bituminous variety, formed at a much later period than the true coals. *Lignite* is another of these coals, and is largely found and used throughout the Rocky Mountain region, and elsewhere. It is also of more recent origin and formation. It is what is known as carbonized wood, and retains the form of the original wood. *Jet* is a more compact form of lignite. *Albertite* is a pitch-like substance found in Albert mine, New Brunswick; *Grahamite* is a similar substance found in West Virginia. *Peat* is partially carbonized vegetable material — imperfect coal — formed in boggy or marshy places, and is of quite recent formation—is forming now, indeed, in some places. Peat is largely used as fuel, where it is found, the peat-bogs of some countries being noted for their extent and value."

VII.

PENNSYLVANIA'S COAL-FIELDS—ANTHRACITE AND BITUMINOUS REGIONS.

"And now, a few general statements in regard to the Pennsylvania coal fields, which we are to visit.

As already stated, the Appalachian coal field is the greatest in the world—notwithstanding the fact that the area given for the Western field is considerably greater than for the Appalachian, as you must have noticed. But it is not alone the extent of a coal field, but also the number and thickness of its beds, and the quality of the coal they contain, which form the basis for estimating its real value; and it is the possession of the greatest beds of the finest coal which places the Appalachian first in the list of the coal-fields of the world. It is the great beds of the upper or northern part of the great Appalachian field which are especially noted; and it is these beds—of both anthracite and bituminous coals—which make Pennsylvania the great coal-producing State of the Union. Let us look first at the bituminous region, as naturally coming before the other —anthracite, as has already been stated, being a changed form of bituminous coal; both having once belonged to the same widespread beds, though now separated by the Allegheny mountains.

One of the first points to be noticed—one which those of last year's party will recollect—is, that the rock *strata* of Western Pennsylvania, in which part of the State the bituminous coal-beds are located, lie flat and evenly, one upon the other, just as they were originally deposited. This condition of things is found in all bituminous coal fields—as indeed over a great part of the globe—and is of much importance to coal-workers, as adding to the ease with which the coal can be mined and taken to the surface. A bituminous coal-bed may cover hundreds of miles in one unbroken extent.

The bituminous coal region covers a large portion of the western part of the State, though the southwestern section contains the most valuable beds. The reason for this difference in the matter of the coal deposits may be readily explained. Those of you who were among the oil wells of this same region, last winter, will also remember what you saw and were told about the dip of the rock *strata*, the coal-beds included—how all the *strata* dip or sink toward the southwest. Suppose, Ned, you see if you can make a picture—a diagram—showing this dip of the rocks. That's it; well done, my boy! I see you have n't forgotten some things learned a year ago, at all events. Here he has shown us this dip, very nicely, so that we can all readily understand the matter. We see the *strata* getting farther below the surface toward the southwest,

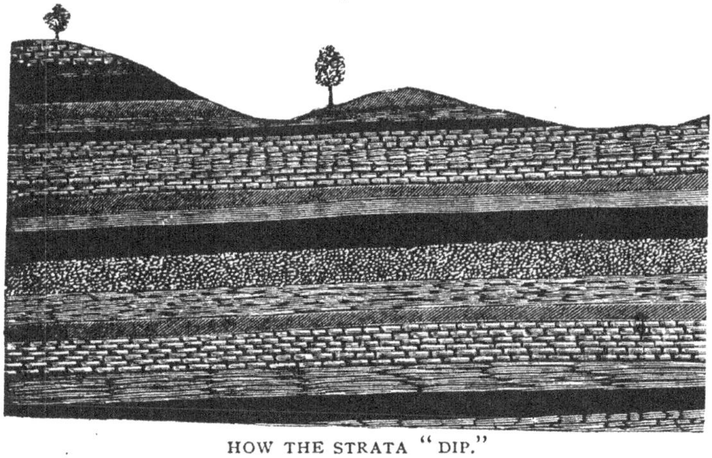

HOW THE STRATA "DIP."

and coming to the surface and 'running out,' toward the northeast. Many of the coal-beds which are found in the southwestern corner of the State have thus 'run out' toward the north, and are not found far beyond Pittsburgh.

There are, altogether, quite a number of coal-beds—some thick, others too thin to be mined—lying at various depths. These beds seem to be in two groups, called the *upper* and *lower coal-measures*, separated by a great thickness of other *strata* containing no coal—and therefore called the *barren measures*. The beds of each group are also separated by various, and often considerable, thicknesses of other strata—limestone, sandstone, slate, shales, etc. It is the beds of the lower group or series—called the *Allegheny series*—which are found and mined north of Pittsburgh, the beds of the upper series having 'run out.' From Pittsburgh southward, the beds of the upper, or *Monongahela series*—beginning with what is known as the Pittsburgh bed, as the lowest—are being mined now; the beds of the lower series being, on account of the dip, far below the surface and, therefore, not likely to be disturbed until the higher beds have been exhausted.

Just here it will be well to state that the beds of rock and coal are not everywhere of the same thickness, but may vary, somewhat, in this particular at points but slightly separated. These slight variations are probably due to the washing about of the deposits of sediment, etc., and also—especially in the case of the coal-beds—to a difference in the depth of the deposits, long ago made. But while these differences may be but slight in any particular locality, there may be very great differences between widely separated parts of the great fields, and especially between the fields themselves. For instance, the thickness of all the rocks of the carboniferous age is 14,570 feet at the Joggins in Nova Scotia, while in Pennsylvania the total thickness is but 9,000 feet. Again, that portion of the *strata* of this age which constitute the coal-measures proper, is over 8,000 feet thick in Nova Scotia ; 4,000 feet at the greatest, in Pennsylvania; and from 1,000 down to 100 in the Mississippi valley fields—showing a much greater depth of deposits in the East than in the West.

The great bed known as the Pittsburgh bed, is one of the most valuable coal-beds in the world—being of great thickness, of the best quality and of great extent. Beginning in the neighborhood of Pittsburgh, where the dip has caused it to 'run out' it covers all the southwestern part of the State, and extends as far east as Cumberland, Maryland—where in the 'Cumberland basin,' it is 12 to 16 feet thick—and southward and westward into West Virginia, Ohio and Kentucky, varying in thickness in different localities. It appears near the top of the great river hills, opposite Pittsburgh, and can be seen as a black belt along the rocky hill-sides for many miles up the Monongahela river. Above the Pittsburgh bed, in the highlands of the extreme southwestern corner of the State, there are other, but less important beds, which are mined at many places.

I think I mentioned a while ago that there were a number of kinds or grades of coal, of each of the two general classes, anthracite and bituminous ; just here it may be well to speak of this matter more fully. While all bituminous coals, as you have learned, are softer than anthracite; they are not all of equal hardness—or softness—by any means; there is a wide range in this particular between the coals of different sections of the same field. The coal of the Monongahela valley, for instance, is quite hard, as compared with coal of the same bed mined in other sections. The coal of the wonderful coke-region is—No, no, no, Miss Question-box, I can't do it now; but you'll learn all about coke, and the coke-region, before you're much older, as we shall visit the cokeries in the course of our travels. As I was about to say, the coal of the coke-region is much softer than that just mentioned; and so, throughout the whole region, there are differences in this particular.

Special names are given to certain kinds of bituminous coal. *Cannel* coal is what I may call a very fine-grained variety, of a dead black color, that is, having but very little, if any, luster; and I may add, just here, that all coal is more or less lustrous, or 'shiny,'—they are diamonds, you know. Cannel coal can be cut with a knife very readily, and if we find any in our travels,you boys may try your jack-knives on it, carving figures or making rings, as I used to do—Oh, no, Bess; I haven't any of them now, I'm sorry to tell you. *Candle* coal was the name really given this variety, on account of the fact that it burned with a candle-like flame, but the Welsh miners pronouncing candle without the *d*, thus, can'el, it came in time to be called cannel coal. *Torbanite* is a variety of cannel coal, from Torbane Hill mines, in Scotland. *Block* coal, found largely in Indiana and neighboring States, and in smaller quantities elsewhere, is of a nature indicated by its name. It is especially valuable to iron-workers, as it does not cake—that is, melt and soften—in burning, and afterward harden or cake over, and therefore will sustain the great weight of iron, etc., resting upon it, in furnaces. Such coals are known as *non-caking* coals; while coals that so cake are known as *caking* or *coking* coals.

By analysis (that is, such a method of examination as will tell us of what a substance is composed—how many separate substances it contains, and the quantity of each of these)—by analysis, the composition of the various grades of coal has been learned. Let us see what this shows, taking the average composition of a number of specimens of each kind, and giving the quantity of each separate substance of which coal is composed—that is, how many parts of each separate substance there are in, say, every 100 parts of coal. Single specimens, from various sections, may differ considerably in their composition; but we will take the average of a large number of specimens. We have, then, the following carbon, $65\frac{18}{100}$ parts; oxygen and hydrogen, $28\frac{36}{100}$ parts; ash, $6\frac{47}{100}$ parts, as the *average* composition of the bituminous coals of the region which we are to visit; other fields will differ from this. Ordinarily, bituminous coal contains from 25 to 35 parts of oxygen and hydrogen gases—the volatile combustible substances, as they are termed. Some coals contain as much as 50 or 60 parts of oxygen and hydrogen. The larger the quantity of these gases, the more combustible, or free-burning, is the coal.

The name bituminous is given to this kind of coal because it burns—by reason of the tar, oil and gas which it gives off when heated—like *bitumen*—an entirely different substance, not contained in the coal at all.

Semi-bituminous coal is, as its name indicates, a half-bituminous variety or grade—that is, much less like the true bituminous coal of which we have been speaking, and approaching more nearly the anthracite varieties. The average composition of the semi-bituminous coal of Pennsylvania is:—carbon, $72\frac{95}{100}$ parts; oxygen and hydrogen, $16\frac{85}{100}$ parts; ash $10\frac{20}{100}$ parts. The semi-bituminous region is situated in the mountainous part of Pennsylvania, between the bituminous and anthracite regions, large quantities of the coal being mined.

The ash of coals, the part left after the coal is burnt—and with which some of you, who help mamma, may think you are entirely too familiar—is made up of the impurities contained in the coal, and which are not consumed. These impurities are, in part, elements contained in the coal plants themselves—silica (mainly), alumina, lime, magnesia, potash, soda, sulphur, oxide of iron, etc., but a very small quantity of some of these being present. Besides, the waters washed into the beds of vegetation earthy matter. Sulphur is the enemy of metal—as you may discover by carrying a bit of it in your pocket, among your coins—and its presence in some coals is in quantity sufficient to unfit them for use in the working of metals.

The anthracite region lies east of the Allegheny mountains, in what may be called the central part of the State. As I have already told you, the area of anthracite beds is very small as compared with the bituminous beds. The Pennsylvania region is by far the largest and most valuable of the anthracite fields of the world. There are thirty beds in the anthracite region of Pennsylvania, though not all present in any one field.

In this region the condition of the rock *strata* is wonderfully different from that noticed in the bituminous region. There, all is order; here, all is confusion. There, the *strata* are unbroken, and lie smoothly one upon the other; here, they are broken, crushed, tilted and jammed together, standing upon edge, turned over on their faces, and doubled up in every imaginable way. While the region beyond the mountains has remained undisturbed, great earthquakes or other forces have at some time, hough long ago, produced the confusion we here find. In some sections, the *strata* have been lifted to great heights; in others, they have been sunken, so as to form great basins, as they are called. The coal-beds, of course, shared in this disturbance, and instead of being found, as in the bituminous regions, spreading out unbrokenly over great areas, they are found in long, narrow, widely separated basins, and at varying and sometimes great distances from the surface. There are three great basins or fields; the Northern, the Middle, and the Southern, the middle field being again divided into Eastern Middle and Western Middle.

The very disturbances which produced this confusion, were largely if not wholly the means of producing the hard anthracite variety of coal—

changing it from the bituminous. The coal-beds being broken up and turned upon edge, in many places, permitted the escape from them of the gases so largely present in bituminous coal, this escape being further helped by the heat and squeezing resulting from the enormous pressure brought to bear upon the beds during the period of disturbance and afterwards. The force of the squeezing to which they were

ROCK STRATA, IN ANTHRACITE REGION.

subjected, and which has left the coal so hard and compact, is shown by the fact that at some places the coal disappears entirely, the other *strata* closing together, while in other places the coal-beds swell or bulge to two or three times their usual thickness! The heat and pressure drove out of the once bituminous coal a large part of its oxygen and hydrogen, leaving it nearly pure carbon—our anthracite coal.

Let us, just here, compare anthracite with the bituminous coal, and learn the extent of the change, as shown by the difference in the quantity of oxygen and hydrogen contained in each. We have already seen that bituminous coal usually contains of oxygen and hydrogen, taken together, from 25 to 35 parts in every 100, and, sometimes, as high as 50 to 60 parts in 100; in anthracite these elements have been driven out until there is left but from 2 to 5 parts in 100, in the purest kinds, and but from 10 to 12 at the most, in the less pure varieties. Quite a difference, you see. It is this difference in the quantity of oxygen and hydrogen contained in the coal which makes the marked difference always noticed in the burning the two kinds of coal—the bituminous kindling quickly and burning very freely and rapidly, and with a mass of bright yellow flame, so cheerful looking on a cold day; and anthracite both kindling and burning slowly, with but a very feeble, bluish flame. A quantity of bituminous coal, burnng so freely and rapidly, is consumed or burned up far more quickly than the slow-burning anthracite.

As in the case of bituminous coals, so with anthracite—there are different qualities or grades of hardness, etc. In the more eastern parts of the region the coal is very hard and flint-like, indeed; this is the purest anthracite. But in the western part of the field it becomes less hard and more free-burning—containing more oxygen and hydrogen ; this is called *semi-anthracite.*

The average composition of several specimens of anthracite and semi-anthracite, showing the number of parts in 100 of each substance, as determined by a number of analyses, is as follows :

Anthracite,	carbon......................89 $\frac{77}{100}$;	oxygen and hydrogen, 3 $\frac{92}{100}$;	ash, 6 $\frac{31}{100}$
Semi-anthracite	" ....82 $\frac{86}{100}$;	" " " 9 $\frac{98}{100}$;	" 7 $\frac{16}{100}$

A single pure specimen may often contain several more parts of carbon ; while the semi-anthracite varieties often run the other way, approaching the composition of the semi-bituminous so closely that it is often hard to distinguish between them. The hardest anthracites contain from 91 to 98 parts of carbon.

The nature of the anthracite region—the *strata* so displaced and broken—makes mining operations very different, in many cases, from those of the bituminous region. But we'll go to the mines, that you may see these operations for yourselves; I've told you, now, all the main facts about coal up to the point of the actual mining or digging of it, and as soon as you can get ready—say, day after to-morrow—we'll be off to the mines—visiting first, I think, for a special reason, the anthracite region. Good bye, then, until we meet to start.

Oh, yes ! hold there, a minute. Here ! come back, every one of you; I almost forgot to tell you to especially prepare yourselves for certain features of our excursion. Provide yourselves with an extra pair, each, of good stout shoes with heavy soles, or with rubber boots, which would be better ; a coal mine is a bad place for fine leather, you'll find. And bring your waterproof coats and cloaks, too. That's all ; good bye."

VIII.

IN THE ANTHRACITE REGION.

"Well, here we are in the anthracite region and ready for sight-seeing. As we cannot afford to lose any time, we'll—Oh, Miss Inquisitive ! just hear her !—what's that big, queer looking house over there ? and what's that odd-looking, big, black hill over yonder ? Two questions at once ! Really, young lady, I'm afraid I shall have to give up if you are going to continue at that rate. However, I can answer you this time. The one is a coal breaker, and—hold, hold ! no more questions about it now—and the other is a culm or rock dump. We'll learn all about both after a while ; but just now we will give our attention to other matters, for, as I was about to say a moment ago, we have no time to waste, and should make our rounds so as to see and learn the most possible in the least time.

We might begin, as I think our question-asker, here, would like to have us, with what we see first, on the surface—the breakers, etc., but these things, as you will find, really come, in order, last instead of first. The breakers are for the preparation of coal for market, but of course it can't be so prepared until it is taken from the beds ; and so all the operations of opening a mine and digging the coal must go before the use of the breaker. We'll follow the natural order of operations and thus reach the breakers and culm dumps by and by.

It being so late, now, we will not go into any of the mines to-day, but making an early start in the morning, we will put in perhaps, the whole day under ground. What! afraid? Oh, but it will never do to come so far to see the coal mines and then have to say, when you get home again, that you only saw them on the outside. It is true that a coal mine is a place full of danger of many kinds; and I never should so much as think of taking you into one, trusting only to my own knowledge of such a place, to take you through it and bring you out again in safety. But we will put ourselves under the care of the 'mine boss,' or some other person who may be given charge of us, and I trust we shall have both an enjoyable and profitable underground experience. We will spend, now, a few hours in visiting some of the works, or collieries, as they are called, and arrange for our tour of to-morrow—at the same time picking up, no doubt, some valuable bits of information by the way.

Suppose we go first to that one over there—the one whose breaker caught Miss Inquisitive's eye so quickly. It seems, viewed from here, to be a shaft or slope col—At it again, I declare! However, your question is a fair one, and one that must be answered, not alone for your benefit, but for that also of the whole party. I will not answer it now, however, further than to say that the openings made in the earth to enable the miners to get at the coal have different names according to the how they are made; 'shaft' and 'slope,' the words just used, being the names of two kinds of such openings. A full explanation would bring up other points connected with the same matter, so I will not undertake it now; but we will devote the evening to a talk about these things.

Well, here we are at the colliery. Ah! look yonder; see that group of grimy, black-faced men, just coming from the doorway. They are the miners, or colliers, who dig the coal from the beds, hundreds of them often working in the same mine. These have just been brought up from the workings, below, having finished their turn, and are now going home. They are drawn up, where the opening into the mines is steep, as here —this being a shaft—in what is called a 'cage,' a certain number only being brought up at a time. I believe ten is the greatest number the law allows to be brought up at one time. Look! there is another group now. Ha, ha, young lady, that's good! How these men would laugh if I should tell them your question. Those little things on the front of the miners' hats and caps, and which 'look so much like litttle coffee-pots,'—ha, ha!—are lamps. Each miner carries his lamp, securely fastened to his hat, the light from which is all he has to enable him to see to work. They have just blown them out upon reaching the surface; but had you been at the bottom of the shaft and have seen these men coming through the long passageways of the mine, I just know you'd have had a question or two about a torchlight procession.

This building here, part of which is placed over the opening into the mines—the shaft—is where the machinery is placed which is used to raise and lower, by means of strong steel wire ropes or cables, the cages carrying the men, as also the mine cars which are brought up full of coal, and are sent down again empty, after another load. Let us take a peep inside. Here is the powerful steam engine and over there is the winding machinery—those big iron cylinders, or drums, as they are called, around which you see the wire cable wrapped or coiled. Tap! tap! tap! goes the gong-bell; that's—were you scared, Bess?—that's to signal the engineer—that is, to tell him what to do, a man at the bottom of the shaft having tapped the bell by pulling a wire which runs all the way down. Now see! the engineer has started the machinery. Do you notice that the cable is winding itself around one drum while it is unwinding and spinning off the other? I hadn't told you yet that the opening or shaft has places for the passage up and down of two cages or mine-cars at once; this is the case, and while the cable winding around this drum is bringing up a full cage, the unwinding of that on the other drum is letting an empty one down. Ah! there it comes—but not a cage with men this time, but a car loaded with great, rough lumps of coal. Away it goes, now, off to the big breaker building—but we will not follow it. Here is an empty car to take its place, you see. The engineer is reversing the machinery, so that the winding and unwinding will change about on the drums this time. Tap! tap! tap! goes the bell again—the man at the bottom tells the man at the top that he is ready. Around go the big drums again—this one unwinding with the empty car; the other, winding and drawing up the full one, this time. And so it goes on all the time. Let us go, now, to yonder colliery, and then around to our home—or what we shall call our home, for the present. We must hurry, too, as it will take us half an hour to reach the colliery; and this, with the time we may spend there, will occupy what time we yet have before starting home.

Quite a tramp that, wasn't it? It was a longer one than I thought it would be, and has taken us much more than half an hour to make it, I find. We will, then, have but little time to spend here. This, I notice, seems to be even a more busy place than the one we have just left. By the way, we have not made our arrangements for to-morrow's trip; this should be done this evening yet, if possible, so that we shall lose no time in the morning with such matters. Wait here a few minutes, watching the miners coming and going, while I call upon the superintendent and learn whether we may be allowed to visit the mines here, to-morrow, or must try some other colliery.

Good news for you, youngsters! We are to be kindly permitted to go down into the workings to-morrow, and will have in charge of us a man familiar with every part of the mine, and who will be instructed to let us take our own time at sight-seeing. Now for home."

IX.

DRIFTS AND TUNNELS.

"Let's see! there's some work laid out for us for this evening, isn't there? Well, gather round the fire here, and we will get at it. Oh! by the way, some of you never have seen an anthracite fire, I believe. Come close, then, and see how anthracite burns, as shown by the fire in this big stove. There is very little flame you see, and that flickering and blueish in color. This notice, also, is where the coal is just kindling, or has been burning but a little while; out here, however, where the coal has been burning for some time, it no longer gives off any flame. No smoke is made either, you notice, a great point of difference between anthracite and bituminous coal fires. The coals lie there bright and red, it is true, but dead-like, it seems, to one acustomed to deal with the free-burning, roaring, bituminous coal fires, especially as seen in the open fire-places or grates. Anthracite, I may state, is not suited for open grates, as is bituminous coal, though sometimes so burned; and so, wherever it is used, stoves for heating as well as stoves for cooking, are in general use. While users of bituminous coal love their roaring, cheerful open fires, those who use the anthracite enjoy some advantages which the others, with their roaring fires, do not. The anthracite fires do not require so much attention as do those of the other kind, in the way of replenishing, a small quantity of anthracite continuing to give out heat long after an equal quantity of the other kind would have burned up and died out. Indeed, many anthracite heating stoves are made as is this one, I see, to be self-feeding, and enough coal is put in at one time to last all day. The coal is put in at the top, filling that cylinder you see running down inside, and as the coal burns below, fresh supplies from the cylinder reach the fire. These fancy colored lights set in the stove are thin plates of mica, which are not affected by the heat. Besides enabling one to see the fire through them, thus saving the opening of doors, they add to the appearance of the stoves, and also give the room a more cheerful appearance than do those made without them. In the matter of cheerfulness, however, the bituminous people, as we might call them, certainly have the advantage. But to partially offset this, again, the anthracite people are free from the clouds of black smoke and soot, which fill the air where bituminous coal is burned.

But now to our work—the promised explanations. Miss Inquisitive wants to—and, of course, you all want to—know something about shafts and slopes. Well, I will try to explain in regard to these things, then, but beginning, however, with some other points which should first be considered.

First, then, I would say that the coal operators—the individuals or companies owning the collieries, secure the coal territory in much the same way that the oil operators secure their territory, as explained last year to some of you—that is, by buying or leasing the land containing the coal-beds. But the coal operator enjoys a great advantage over the oil operator in the certainty of finding what he is after. The oil man can only hope, or, at the most, expect, to find oil in the land he has bought or leased, since there cannot be any certainty of its existence there; but the extent of the coal-beds is so well defined, in general, that in securing territory for mining purposes, the operators know that their land contains coal. But the expense of opening up a mine and of taking out the coal, far exceeds that of sinking an oil well and securing the oil—if there happens to be any at that point to secure. It requires a large amount of money for the erection of the necessary buildings, the purchase of machinery, coal cars, etc., the opening of the mine by a shaft, or in some other way, and to pay the large number of miners and other workmen that must always be employed. In the case of oil operations, the expense after the well is completed is but small—the well perhaps flowing of itself at a wonderful rate for a long time, and after this ceases requires but one man to keep the pump going. But in mining, while the expense of opening the mine and getting everything ready to handle the coal produced is very great, indeed, yet the expense does not end here, by any means. The coal does not flow out of itself, hundreds or thousands of dollars worth of it a day when the bed is reached; oh, no; it can only be even then secured at a great cost of time, labor and money, as we shall see. As large capital is thus required, the business of mining is, for the most part, in the hands of wealthy companies, who own, or have leased, large tracts of coal land.

The cost of opening and operating a colliery depends upon several things, but more largely than anything else upon the position of the coal-bed or beds to be opened up, requiring this or that kind of an opening to be made to reach the coal to be mined. You see how rough and rugged the country is, and you have been shown too, how the rock and coal-beds are twisted and wrinkled and jumbled up; in some places being flat, in others standing on edge, or doubled back upon themselves, and so on.. All this affects the cost of operating a colliery.

Let us see, then, how mines are opened up—which will soon bring us to the answer to Miss Inquisitive's question. There are four ways of opening a mine, each of which we will examine, beginning with the one the least difficult and expensive. With some of these we may not meet in this locality, however, perhaps not at all, during our brief stay in the anthracite region. This being the case, the present explanation should be listened to the more carefully.

The method to be employed in the opening of a mine is not always a matter of choice on the part of the owners, but is indicated by the position of the coal-bed or beds in relation to the surface of the land embraced within the territory of the company. The conditions may vary so much in adjoining properties as to require a different plan of operations in each. The simplest and cheapest form of opening, and the one employed wherever possible, is what is called the *drift*. Drift mining, however, is possible only where the coal-bed is flat, or nearly so, and 'crops out' along a hillside. What do I mean by 'out crop?' Why, the 'out crop' of a coal-bed, or of any rock bed, is where it appears along the face or side of a hill. It—Here, Nell, hand me that piece of paper; Ned, lend me your pencil. Now, I'll show you what I mean. Here we have a hill. Now, I'll make some rock *strata*, and just here I'll make a coal-bed—making it 'black as coal,' as the boys say. Now, where the rock and coal-beds show on the side of the hill, as you see they do, they

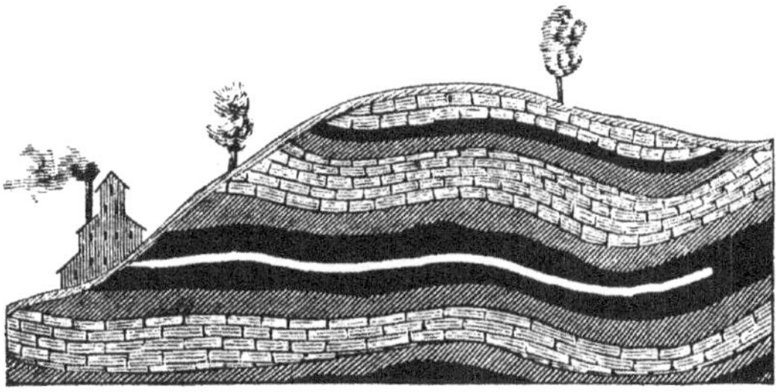

"DRIFT" MINE.

are said to 'crop out.' They may not always be seen thus cropping out, as they are generally covered with a thin coat of soil where the hill is not very steep; but along steep hillsides, where the soil cannot gather to hide them, they can be seen—that is, their edges—just as you may have seen them in cuts and quarries.

Since I have this picture, I may as well use it to explain more clearly drift mining. Here, then, on the side of the hill, where the coal crops out—here at the left, Bess—the opening would be made right into the coal-bed, or 'seam,' as it is commonly called, and carried back along it into the hill. I'll rub out the pencil mark, along here, and the white streak made in our coal-bed will show how far back the mining has been carried on. Oh, no; the mining is not carried on along just a single opening, but branches out from it, right and left—as we shall see to-morrow. The coal in such a mine can be hauled right out, in little cars, by horses or mules, without the use of the expensive winding machinery we saw in operation this afternoon; tracks upon which the cars run being laid on the bottom or floor of the mine. Right here we will put the breaker, which is always as near the mouth of the mine as the nature of the ground will permit. There! now we have a very fair representation of a drift mine. Much of the earlier mining in the anthracite region was done in this way; but not much of it is now done, as nearly all the coal that could be so reached has already been taken out.

Our picture does not, of course, show us what a mine is like inside; this we will see to-morrow. I will say, here, however, that the opening into a mine, as well as all the other openings throughout the mine—the gangways, airways, breasts—Hold! hold! one at a time, and I'll try to answer you. No; I won't, either. We'll leave these questions unanswered for the present and go straight ahead with the matter already in hand. To a greater or less extent, in different localities, all important openings made in the coal seams, by mining the coal, must be 'timbered'—that is, have great numbers of strong posts set up and beams otherwise placed, to hold up the roof or sides. In some mines the roof may be of such a nature as to need no propping in some places; but as a rule, timbering is necessary. To leave the roof unsupported, would be

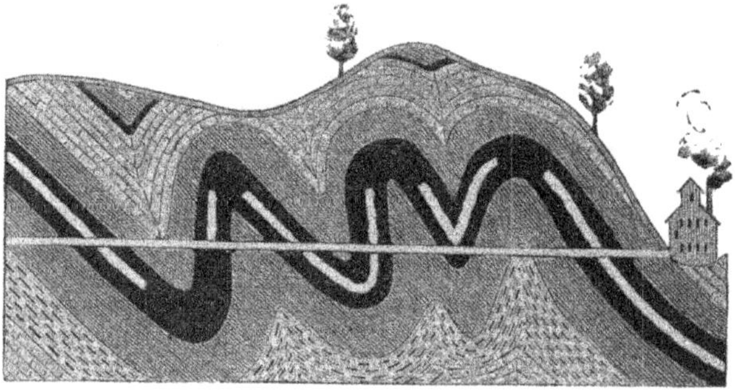

"TUNNEL" MINE.

to have it constantly falling down, killing or injuring the miners and blocking up the mines. In such an opening as this I have shown, as in other similar passages, a row of heavy posts lines each side, while heavy beams rest upon the upper ends, running 'crossways' of the opening. Other lighter timbers or planks are placed behind the posts and on top of the cross-beams, (running lengthwise, that is,) and thus the roof and sides are kept in place.

Now for another kind of opening—the tunnel. As in drift mining, the opening is made from the side of the hill, but it is made not in the coal seam, but in the rocks, through which it is carried in order to reach the coal, which is so situated that a drift opening cannot be made. Let me make another picture in order to show you how the coal-bed lies where a tunnel, instead of a drift, opening must be made. The coal, you see, does not lie flat, or nearly so, as in the drift mine I have just pictured, but runs—that's it, exactly, Nell—'every which way.' Now in such a case, an opening would be made here at the right and carried in through the rocks until the coal seam would be struck—just here.

From this point mining would be carried on, an opening being made through the coal and the work extended to each side. The tunnel would also be carried further ahead until the coal would again be struck —here—and mining would then be commenced at this point. The tunnel would again be driven forward, and at each point where it struck the coal, mining operations would be begun—the coal being hauled out through the tunnel to the breaker. These white spaces in the coal seam show where the coal has been taken out, as in this other picture—no mining having been done in the solid black sections appearing between the ends of the white spaces. Where the coal-beds lie about as here shown, tunneling is the best way to reach them.

In neither drift nor tunnel workings is it necessary to use machinery in hauling the coal from the mines. We come now, however, to openings of another kind—those through which the coal can be taken out only by the use of machinery. Such openings are either *slopes* or *shafts*. And now for the answer to Miss Inquisitive's question; poor girl! how long she has been waiting for it!"

X.

"SLOPES" AND "SHAFTS."

"Slopes and shafts are necessary where the coal-beds are found, not in the hills, but deeper; that is, at a considerable depth beneath the general surface of the land, and can only be reached by sinking an opening to them. The points of difference between a slope and a shaft we will now consider.

A *slope* opening is made where a coal-bed, while lying, in general, some distance below the surface, is, by reason of an upward turn of the *strata*, brought to the surface, or very close to it, at least, at some point on the company's property. It is at this point that the opening—the slope—is made. It is made in the coal-rocks having to be passed through for a short distance at the top, in some instances, however, and runs directly down the sloping seam, the name *slope* being, therefore, exactly descriptive of its general character. One or more railway tracks are laid on the bottom of the slope, and the coal as it is mined, far out on either side, is put into cars and drawn to the surface by hoisting, or winding machinery, as it is called, such as we have already seen. I will make a picture of a slope, also; there, that will do. This white streak represents the slope, of course. Away down here, you see, I have made the coal-bed nearly flat, as it may be for some distance before again rising.

The pitch or slant of a slope will depend, of course, upon the steepness of the dip of the coal seam down which it runs. In some mines the slope is very steep, indeed, while in others it has but a slight pitch, and the miners can readily walk up and down it. How high? Usually about seven feet between the railway tracks and the roof-timbers; sometimes more, however. Slopes, as a rule, require most substantial timbering—sides and roof—throughout. They are, too, divided into three or more compartments throughout their whole length. Two of these compartments have tracks laid in them, being used for the drawing up of the coal, etc. The third compartment is occupied, in part, by the pump-pipes used in keeping the mine free from water, and by other fixtures, in some instances. It is used at some collieries as an airway, also.

We now come to the fourth kind of opening, the *shaft*. Shaft collieries are the most expensive of all. But there is no help for it; in many places if the coal is to be taken out at all, it must be through a shaft.

"SLOPE" MINE.

Shaft openings, then, are made where the coal-beds lie beneath the surface, and without coming close to it at any point, as it does where slopes are made. Much of the coal is thus situated, lying at different depths, in different localities. A shaft differs from a slope in that it is sunk vertically*—or, as we commonly say, straight down instead of slanting; also in that it is, of necessity, sunk through the rock *strata*, until the coal seam is reached, instead of in the coal seam itself. I will make a picture of such a colliery, also. Here is the coal-bed, lying, say 500 feet from the surface. Here is the shaft, running down through the rocks overlying the coal. When the shaft has been sunk down into the coal, openings are made in the seam at each side, as in the other instances, and mining is begun, the coal being lifted from the mine through the shaft by

* In the anthracite region, "shaft" is understood to mean such a *vertical* opening. In some other localities the name is given to a *sloping* opening similar, in general, to the "slope" of the anthracite region (as already described), but sunk of necessity *entirely in rock* to the deep-lying coal, instead of down the sloping coal seam itself.

the heavy and powerful machinery at the top. These white spaces, as before, show the extent of the 'workings' where the coal has been mined. A shaft may be made to reach several beds, at different depths, of course, and mining be done in each.

How large? Oh, of various sizes. They are from 10 to 12 feet wide, but vary greatly in length, according to the number of compartments it is desired to have. For two compartments they are from 16 to 20 feet wide; for three compartments, 22 to 26 feet; 30 to 38 for four; and 44 to 52 for the extra large ones having six compartments; only a few, however, are made of these large sizes. Where two hoisting compartments are used, one is for an up, and the other a down cage, as you already know, the same openings being used for the hoisting of both men and coal. As one of the laws of the state, for the protection of miners, forbids the hoisting or lowering of more than ten men at a time, and as no coal

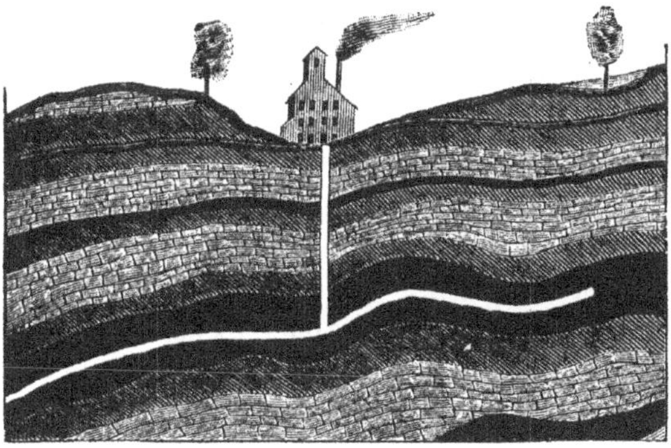

"SHAFT" MINE.

may be raised during the time of the passage of the men, up or down, a loss of one or two hours of coal hoisting every day is occasioned by the coming and going of the miners and other workmen. This loss is prevented in the case of large shafts with four or more compartments; one of them being used for the men and two for coal exclusively. Others are used as airways, etc. The quantity of coal, by the way, which some of the largest collieries put out, is very large—a thousand or more tons per day. To do this, the cages are kept fairly flying up and down the shaft as rapidly as an express train! a wonderful contrast to the methods employed and the results obtained, by our grandfathers.

Dig them? Oh, no, excepting at the top where soil and clay cover the rocks. Shafts, slopes and all the openings I have mentioned are made by blasting. Deep holes, an inch or two in diameter, are sunk into the rock by means of drills, (bars of steel, with an edge at one end, or machines which are made to bore into the rock,) and these holes are filled with powder which is then exploded, blowing loose the rock or coal. In shaft-sinking a method of deep drilling has been employed successfully. A large number of holes, distributed over the surface embraced within the shaft, are drilled by means of machinery to a depth of 200 or 300 feet. and are then filled with sand. When a shot is to be made, five or six feet of the sand is taken from each hole, and the space left is filled with powder, which is exploded at the same instant in all the holes. When the broken rock has been removed, another shot is fired, and so on to the bottom of the holes. The depth of shafts, here, is from 400 to 600 feet; 800 in some other places; while a depth of nearly 1,600 feet has been found necessary in some localities. These depths, you will understand, represent the distance of the coal-beds below the surface, in different places.

IN YE OLDEN TIME.

The opening of mines, especially by means of large shafts, is very slow and expensive work, 200 to 300 feet per year being ordinarily the depth made. But besides the main opening, a law of the state makes it necessary that a second opening he made at a distance of not less than 150 feet from the first. This is for the protection of the miners, to give them a double chance to escape in case of accidents, explosions, fires, etc., in the mine. These second openings, however, are smaller than the main ones. Some shafts, because of the nature of the rocks through which they pass, require complete timbering. Inside this are placed, running from top to bottom, the strong wooden 'guides' against which the sides of the cages rest, sliding up and down. Where complete timbering is not required, timbers or beams are placed across the opening, the ends resting in niches or pockets cut into the walls of rock, and to these the guides for the cages are made fast.

You notice that I have made the white spaces representing the workings in the heart of the coal seam; perhaps I should explain in regard to the location of the main openings, and of others inside the mines. The beds which are now being mined are the best in the anthracite region—the greatest in thickness. There are about thirty coal-beds or seams in the anthracite region, although only part of them may be found in any one of the great basins. These beds range in thickness from the very light or shallow ones, not now made use of, to the great Mammoth bed, which at different points is from 30 to 50 and even 60 feet thick. At some points this bed is completely doubled back upon itself, giving 100 feet of coal in one thickness. But the coal-beds are very seldom composed entirely of coal, but are divided up into parts of various thickness by seams or 'partings' of dirt, slate, bone,' as it is called, etc. These partings, which in thickness are from an inch or less, in some instances, to a foot or more, in others, make the refuse material, or 'gob,' in mining, and give no direct return in money for the expense incurred in digging and handling it. However, a heavy parting in a thick coal-bed is often a very great help to mining operations. The separate parts of a coal-bed thus divided, are called 'benches' and it is much more easy to mine out one bench after another, between the partings, than it would be to take out a great thickness of coal without such partings.

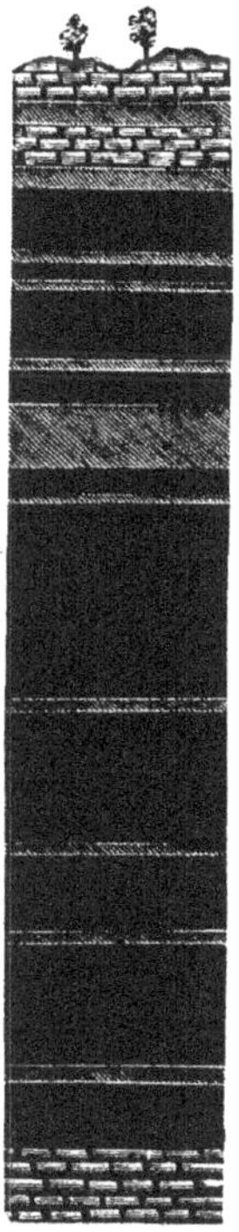

COAL-BED, WITH "PARTINGS."

These partings usually make up, as taken together, a considerable portion of the thickness of a coal-bed, and often much of the coal in a bed may be left untouched at the bottom or the top, as the case may be, because it is so divided by these 'dirt' seams as to make it unprofitable to mine it. In driving the main openings in the very thick beds, they may be made at the bottom, or higher, as mining experience shows to be the most advantageous under the circumstances surrounding each case—the coal lying higher being taken out, later, as the work goes on, opening out at each side. Where the beds are not too deep, the whole thickness of good coal from floor to roof is taken out at the one operation, as is commonly the case in the bituminous regions.

Now I have explained the various methods of opening a mine, so far as the making of the main opening, from which others branch, and through which all the coal is brought to the surface, is concerned. But it is through the other openings, gangways, etc., alluded to a while ago, that the great body of the coal is reached and taken out. As it is yet early, let us learn something about these other openings and how the coal is mined, so as to become somewhat familiar, in a sense, with these matters, before seeing them to-morrow. The less time spent in explanations at each point, the more we can see in a certain time, you know."

XI.

PLAN OF "WORKINGS"—GANGWAYS, SHUTES, HEADINGS, BREASTS, ETC.

" When the opening, of whatever kind, has been carried far enough for the purpose, the work of extending the mine, on each side of the opening, is begun. The principal opening of those within the mine, and the one first made, after the shaft or slope has been carried far enough, is the *gangway*. It is carried off from the shaft or slope on each side, nearly level, and extends far out through the coal. It is the passageway for the bringing of all the coal mined throughout a large part of the seam to the shaft or slope, to be raised to the surface. It is the main street, we may say, of the mine, into which all the travel and traffic of the side and cross streets is carried. It must, therefore, be made wide enough, high enough and level enough, to accommodate this travel and traffic, and must be well protected, where necessary, from being blockaded by the falling of the roof or sides. It is, then, usually made from 9 to 10 feet high, and from 10 to 14 feet wide, at the bottom, or floor, of the mine. The sides slope inward toward the top, which is, then, of course, less wide than the bottom, or from 9 to 10 feet. Timbering in some gangways is needed on both sides and at the top; in others, especially where the coal-bed dips steeply, the upper side only, or the upper side and roof, will need such support. But whenever and however used, the timbering must be heavy and well built in, as it is of the greatest importance, as you can see, to keep the gangway from being obstructed by falls of coal or slate, which would, if they occurred, stop all travel through the gangway and stop also a great part of the work, inside and outside, until the track should be cleared for the passage of the mine cars, again. For some distance on each side of the slope, as we will say it is, the gangway is made wider, to give room for the passage of the cars, loaded and empty, which center here from all parts of the mine—the loaded ones to be raised, the empty ones as sent down to be carried back into the mine, to be again filled. This wider part is called the 'turn-out.' Double-track gangways are sometimes made, which are, of course, wider than the kind I am describing—single-tracks—which is more generally employed.

The gangway is made, or driven, we will say, by blasting the coal—as is the case with other passageways. Either the upper part—a 'bench' it may be, above a dirt or slate parting—is first removed, by blasting with powder, again and again, and the lower part afterward blown up; or, the lower part is first blasted out for a short distance, and the upper part is then blown down.

Ah, Miss Inquisitive, here you come again. Want to know how long a gangway is, eh? Well, that's a very innocent question, to be sure, and one which I will answer for you, straightway. Gangways vary in length, and for more than one reason. First, the length must be limited by the extent of the territory owned or leased by the company operating the colliery. They must not, of course, be carried across the line into the coal of a neighboring company, as—one at a time! please, one at a time! even if you do all ask the same question. Yes, indeed, the company knows perfectly well just when it has reached the limit of its land, as measured on the surface, though the surface may be hundreds of feet above. How? you all ask. Well, in the seam way, in general, that measurements are made and distances found at the surface—by surveying. Underground surveying? Yes, indeed; and it is so successfully and accurately done that the limits of the company's coal-beds are found to a nicety. But to proceed: In the second place, a gangway is limited in length by the nature of the roof and sides, and the expense necessary to keep them up, and the track clear. Where the roof is of such a nature that but little if any trouble and expense is incurred in keeping a gangway open, it may be carried several miles from the shaft or other opening; but in other cases, where the roof is so poor as to be a continual source of annoyance, requiring much outlay for repairs, in order to keep a gangway open, throughout a great length, the distance to which the gangway is carried is much less than under different circumstances—perhaps not more than a mile, if so far. In such a case, a new slope or other opening may be made, and new gangways started, as being less expensive, and otherwise preferable to keeping a long gangway open under such difficulties.

"DRIVING" A GANGWAY.

But we are 'ahead of developemnts,' as the oil mên say of the 'wild-cat' operations. Long before the gangway has reached such distances as those mentioned, indeed, before it has been carried more than a comparatively short distance from the slope, other openings are being made. You all know how necessary it is for you to have air, fresh air, and plenty of it. Well, the miners are just like you and everybody else, in this respect. They must have fresh air, and enough of it. But hundreds of feet below the ground fresh air is not to be had for the asking, or the taking—as here above ground; in fact, it is not to be had at all, we may say, under ordinary circumstances. To overcome this difficulty, and to make the air of mines such that men can work in comfort, so far as breathing is concerned, fresh air is sent down the mine from the top, and carried in *airways* to every part of the mine. The airway is not nearly so large as the gangway. One of the mine laws of the state says that it must not be less than 20 square feet in area, which would be 4 feet one way and 5 feet the other; but they are commonly made even larger than this, I believe, by considerate and careful mine owners. The air is forced down in the mine from the top by immense fans—wheels with blades or wings to gather the air. At the bottom, it follows the airways made for it. The airway is usually made alongside the gangway, and rather at its top, and on the upper side, when the coal seam dips. As the air is to be carried to the farther end, and then made to flow back through all the parts of the mine where the miners are at work, it is necessary to confine it, to prevent it from mingling with the air in the gangway, or other openings; a certain quantity of it, only, being allowed to get into the gangway, through holes made for the purpose. To thus confine the air, the side of the gangway is made air-tight by planking it. This plank partition is called a 'brattice.' The airway is carried back along each gangway, as the gangway is extended.

To get at the main body of the coal, other passages and openings are made, some running off from the gangway at a right angle to it, and others connecting these again, running parallel with the gangway. To help you understand the arrangement of these openings, I think I would better make you a picture of the inside of a mine, showing these openings. While I am drawing it, you may talk, or sing, or play, as you please.

Here, now, is our picture—representing a slope colliery, and one in which work has been carried on for quite a long while, already. This is the slope—running down here, the center of the workings; you see I have marked it, as well as some of the other openings. This is the surface where the coal crops out—up here, at the right; and the breaker and other buildings would be placed just here. Over here, and running parallel with the slope, from the surface, is an airway—smaller than the slope you see. These large white spaces, here, one on either side of the slope, are old workings where the coal has been taken out, the work now being carried on farther down the seam.

Of this matter of workings, let me speak more fully. Mining carried on as I have here shown it, is called 'lift'-mining. When the slope has been carried down, say, 100 yards from the surface, (it may be from 75 to 125 yards in different collieries,) a gangway is opened out on each side, and the coal is taken out on the upper side of the gangway, toward the top; this is called the *first lift*. I have shown the first lift here, as having

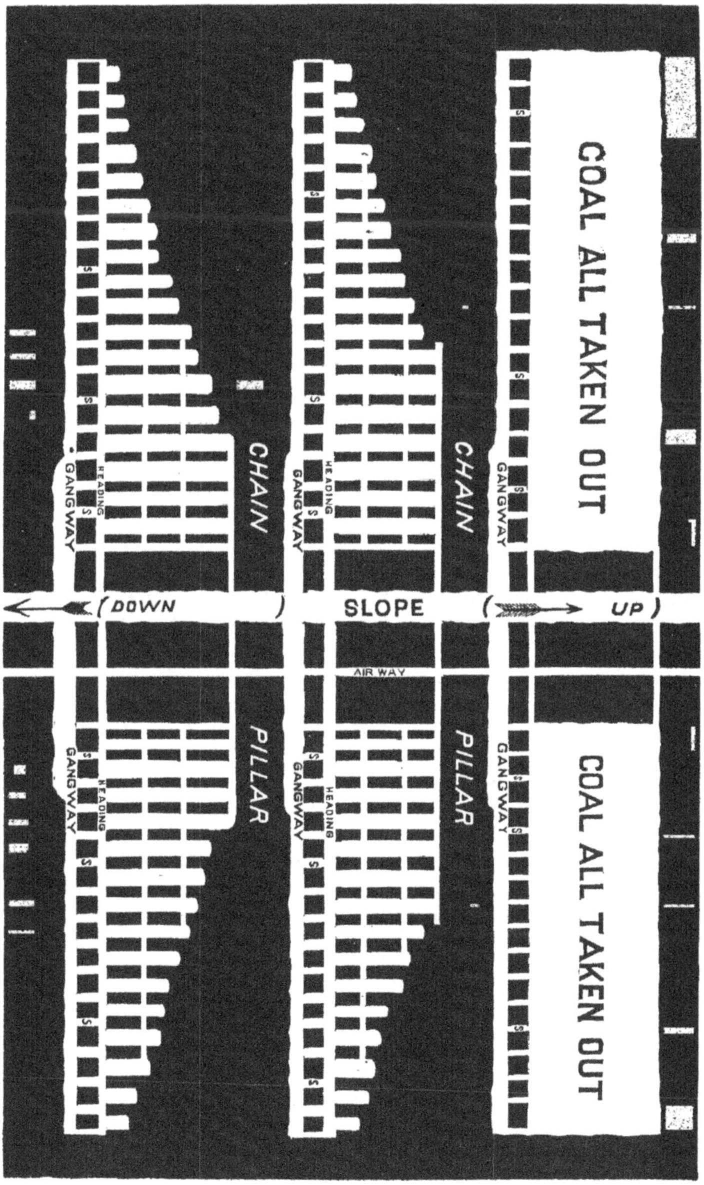

INTERIOR VIEW OF COAL MINE.

been finished, the coal all having been taken out, as you see, excepting this row of nearly square blocks, along here. These blocks are called *stumps—gangway stumps*—and are left to hold up the roof of the mine, in order to prevent a 'crush' or 'squeeze.'—No, not now; I'll tell you about these things at another time. Here is the gangway, you see, between the stumps and this thick strip of coal on the lower side. The gangway, you see, extends on either side, to the limit of my picture, which we will suppose is also the limit of the company's property in these directions.

Here, again, another 100 yards lower, we find other workings—the *second lift*. The thick strip of coal left unmined between the first and second lifts, and which, I may add, is always left between any two lifts or workings, is called the *chain pillar*. It is of considerable thickness, and is left to firmly support the roof; if taken out a 'crush' or 'squeeze' would follow, perhaps destroying the whole or a very great portion, at least, of the workings throughout the mine. You notice, too, that similar strips of coal, though not so thick as the chain pillars, are left standing along the sides of the slope, between it and the workings. These are called *slope pillars*, and protect the slope from being closed by the falling of the roof, assisting the chain pillars in preventing serious accidents of this kind. A strip of coal is also left above the first lift, next to the surface, as you see.

Let us examine the workings here, in the second lift. First, here is the gangway, at the lower side. Ha! ha! you couldn't stand it any longer, young lady; could you? I know you must be fairly aching to ask me about those white and black patches, as you call them. Well, I'll satisfy your curiosity, at once. These are the other side openings to which I alluded a few moments ago. The gangway having been carried back some distance from the slope, in opening up a new lift, the mining of the coal on the upper side is begun—the gangway, at the same time being driven still further back. When I say 'upper side,' I mean upper side; in this picture, the side to the right. Always keep in mind the fact that we are now speaking, not of a flat coal-bed, but of a tilted or pitching one, the slope coming down the seam, like a road down a hill, and the gangways running the other way, like roads running along the side of the hill. The gangway is made at the lower side of a lift to save labor in handling the coal, after it is mined; the coal being taken out above the gangway—further up the hill, we will say—it is allowed to slide down along the floor of the workings into the cars standing in the gangway to receive it. If the gangway were made at the upper side, with the workings below, all the coal, as mined, would have to be raised up, instead of sliding down of itself, as now, to be put into the gangway cars. A vast amount of labor is saved, then, you see, by following the plan first described, and which, I think, you all understand.

All the coal on the upper side of the gangways is not taken out, as you can see by the picture. Strips of coal, as pillars or stumps, must be left at certain distances throughout a mine to keep up the roof. The gangway itself must be so protected; and so a considerable thickness of coal, 6 to 12 yards, is left at the upper side. The real workings, then, do not begin at the gangway, but at another smaller opening, running parallel with the gangway, at the upper side of the strip of coal to be left standing along the gangway. This opening is the *heading*—here—and is usually 4 to 6 feet wide and 6 feet high. From this are opened up the *breasts* or *rooms* in which the real mining is done. But as the coal mined is to go to the cars in the gangway—sliding down, as I have stated—there must be openings, corresponding with each breast, made in the strip of coal between the gangway and heading, through which the loose coal can pass. These openings are called *shutes*, and are made 4 to 6 feet wide and 4 feet high. Here they are, these narrow openings which connect the gangway and heading, and in some of which I have made an S—for shute. By means of a platform built in the lower end of the shute, to a height equal to that of the mine cars, the coal slides down the shute and drops into the car standing on the gangway track. In very steep-dipping beds, the loose coal comes tumbling down at a great rate, while in those where the dip is comparatively slight, it must be pushed along. These shutes divide the strip of coal left standing into a number of blocks, as you see; these are the gangway stumps—the same as you saw yet standing in the first lift. I was not careful, enough, I notice, in making my picture. Some of the shutes appear wider than others, and the same is true of the stumps, as also of some of the other openings and pillars. We can correct this in our minds, however.

Now for the breasts—the real mining places. These white spaces above the heading represent the breasts, the black strips between being the pillars left standing to support the roof. Breasts are commonly about 8 yards wide, but sometimes, are made 12 yards wide—the width varying in different mines, depending upon the nature, to some extent, of the roof. The pillars between the breasts are usually equal in thickness to the width of the breasts. Where the roof is poor, they are not so thick as in other cases, but are more numerous, the rooms being less wide. These pillars, however, are not to remain permanently, the coal in them thus being lost; but when all the breasts have been mined out, up to the chain pillar, like these nearest the slope, the pillars are removed—'robbed out'—beginning at the farther end, and coming toward the slope, as has been done in this first lift, you see.

These first breasts or rooms, nearest the slope, have been worked entirely out, that is, they have been carried up to the point where the coal is to be left as the chain pillar. Farther on, you see, they are not completed, having been begun later, as the gangway was driven farther out. Away out at the farther end of the gangway, they are but fairly started. These narrow openings which connect the breasts, or rooms, and which cut the pillars between them, are *cross-headings*. They are made for the passage of the currents of fresh air, from breast to breast; as it is necessary, and required by law, that a good supply of fresh air be kept constantly passing through every breast or room in which work is being done, close up to the end of the breast, where the miners are working.

Down here, now, 100 yards farther, is another lift—the third. Not quite so much work has been done here, as in the second, you see. Work may be carried on in three or four lifts at the same time, all the coal going up the slope, landings for the cars being made at each lift. The arrangement of gangways, breasts, headings, etc., is the same here, as in the other lift, you notice; and so it would be found in other lifts farther down. In shaft-mining, where several coal seams are passed through, two or three or four of them may be worked at the same time, just like so many lifts in a slope mine.

Where the seam dips steeply, as we have supposed it does in the mine which we have pictured, the mining operations are carried on as

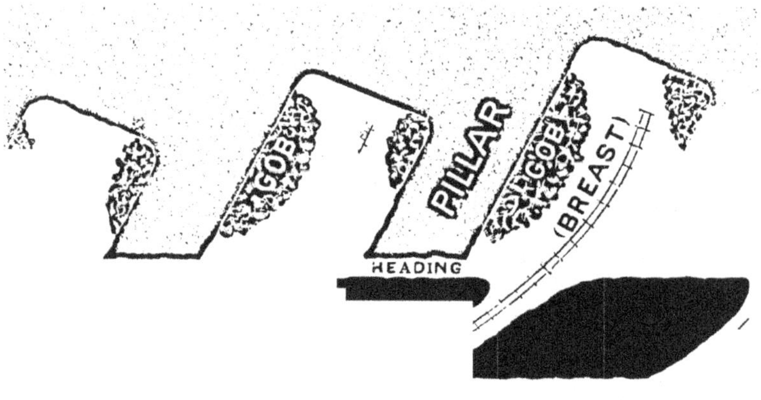

WAGON BREASTS—IN FLAT WORKINGS.

here shown,—the work being done on the upper side of the gangway, the loose coal sliding down to the gangway. Where the dip or pitch is not very great, another method is employed. As the coal cannot slide down, the dip being too slight, it is sent down from the breasts to the gangway, as fast as mined, in *buggies*—Hold! hold! I'll explain, straightway. Didn't expect to hear of buggies in a coal mine, did you? Well, there are buggies in the coal mines, and many of them, but rather different from the buggies you have in mind. A coal-mine buggy is an iron box, on small wheels—you'll see some of them to-morrow, perhaps. They are smaller than the cars which run in the gangway, and made as light as possible, to make it more easy to move them about. The miners pull them from the gangway, through the passages made, up into the breasts, load them with the coal mined, and then run them down to the gangway, where the coal is dumped on to a platform and then shoveled into the mine cars, or, it may be, dumped directly into the cars from the buggy. Breasts from which the coal is sent to the gangway thus, in buggies, are called 'buggy breasts.'

Then there is another method, employed where the coal seam lies flat or level, or very nearly so,—as is largely the case in the bituminous coal region. In such mines there is no upper and lower sides along the gangways, you can understand; and so breasts may be opened out on both sides, instead of on but the one. From these breasts the coal is taken

"BEARING IN."

out in small mine-cars, or *wagons*, as they are here called. A track—a switch from the main track, in the gangway—is laid right into the breasts, and the cars are hauled in, loaded and hauled out again. The breasts may not always be opened straight off the gangway; but in order to bring the grade right, or for some other reason, may sometimes run off —how shall I say it? Here! I will make a picture of it. These are the breasts, you know; here are the tracks into them; and here I'll make a wagon in this breast. As the miners loosen the coal they load it into the wagons, beside them, throwing the refuse, or 'gob,' to one side, against the pillar, where it is out of the way.

The coal, I should have told you earlier, is not found in pieces or 'lumps,' as we buy it. It is all one mass, and must be broken up to get it out. In this, the anthracite region, this breaking and loosening of the coal is accomplished by means of blasting,—drilling holes into the solid bed, and exploding powder in these holes. This method is made necessary by reason of the very great hardness of the coal, digging or picking it, as in the bituminous region, being out of the question. In the bituminous region, where the coal is not so hard, you know, the miner gets down upon the level, or nearly level, floor of the mine, in a half-lying position, and with his short miner's pick— and with only the light from his little lamp to guide his operations — digs back under the face of the coal bed, before him, as far as he can reach — three or four feet; this he does across the full width of his breast or room. It is slow work, as but a little of the hard, black rock before him flies off at each blow,— as dust or very small bits,— and thousands of blows must be struck. This work is called 'bearing in.' The sharp bits of flying coal strike him in the face; the dust fills his eyes, and blackens his hands and face; and his back may ache; but he must keep steadily on until he has carried the opening under the coal clean across his room. Not a pleasant occupation, that of a coal miner, is it? Next, he must cut, or pick, if you please, a similar opening, to the same depth, at one side of the coal, along the pillar left between him and his neighbor of the next room. These openings are about a foot wide at the start, but run to a point at the farther end, in the coal. Having made these openings, to give the coal a chance to fall, he then drives wedges between the top of the coal and the overlying slate or other *strata*,— and down comes the huge block of coal, for which he has been so long and so steadily picking away. A little blasting is sometimes done in the bituminous region.

Now, to bed."

XII.

"DOWN IN A COAL MINE."

"Whew! What a ride! Rope break? Well, let us trust that it won't; but if it should, as there are safety catches provided for just such emergencies, we should probably come to a sudden stand-still, unharmed. Ah! here we are at the third lift. Out with you! Here! this side, out of the way of the cars; you see, this is a busy place, with the cars coming and going. Here comes a train of mine cars. One, two, three, four, five mules pulling it. How odd they look, tramping along one behind the other, single file, the leader with his lamp fastened to his breast! Proud of it? Who knows but he is, as much so as his owner is of the diamond which flashes upon *his* breast. Sometimes the lamp is fastened to the leader's head. See how well trained they are, how intelligent, indeed; how nicely they turned off upon the proper switch—pulled the cars up as far as they could—stepped off the track one after the other, while the driver unhooked the train—wheeled and went over upon that other track, in front of the empty cars—and marched off again, when hooked to them, all without a word! Never let me hear one of you making fun of a mule, after this. Why, they would hardly know how to get along in coal mines without mules, many thousands of them being thus employed. Some horses are used, but mules are much more desirable. Small locomotives are also used in some instances, but there are so many serious objections to their use that the mules are not likely to soon be displaced. Here are their stables just back of us, being placed close to the slope that the mules may be quickly taken out of the mine in case of fire or other accident. The stables, you see, are well lighted, and are quite clean and dry, and comfortable in every respect, and the mules may be as well contented, down here, as are their friends and relations outside. Here, too, is the blacksmith shop, where the mules come when they want a pair of new shoes—or two pairs, it may be.

Ah! here are our lamps—one for each. The boys can easily arrange theirs, but I don't know how it will be with the girls, as the lamps were never made for such hats, or for hoods and bonnets. By the way, I am reminded that I once read—and saw a picture, too, I think—of the employment of women and young girls in the coal mines of one of the foreign countries. Whole families lived in the mines, many persons going outside but once in a long, long time; just think of that! Their condition was most wretched, too, even the women and girls being barefoot, and, indeed, half naked,—only a coarse skirt being worn, in many cases. The husbands and fathers mined the coal, while the women and larger girls drew it from the rooms in wagons, (which were fastened to them by means of chains around their naked waists!) often crawling upon hands and knees, over the rough stony floor, where the coal seam was so thin as to have left but a small space between roof and floor. Dreadful!—dreadful, indeed! But there is nothing of this kind in our day, I believe; so you needn't look as if you would like to get out of here as soon as possible, you girls.

Well, if you all have your lamps arranged, we'll be off. We will follow the gangway, some distance, first. Here comes another train, the driver's lamp, and that of the leader, shining like twin stars in the darkness. Away ahead yonder are the lights bobbing about, as workmen move from place to place. There! see that one suddenly,—why, there's another, and another, and another,—and just look at them! It is a gang of miners coming from the workings, and hid from us by the wall of the gangway, until they suddenly appear, coming from a shute. An odd and pretty sight, indeed! the lamps looking like so many stars. As they come nearer, however, the forms of the men will begin to be seen, and the illusion of stars will disappear. I suppose our party presents the same appearance to them.

We will turn off, here, and go up into the workings, through the shute ;—all the rooms lying off from the gangway, this far out, have been worked out, our guide tells me. These shutes are used as manways, as well as for the passage of coal. Pretty steep, youngsters, and hard climbing, but I think we can make it. Ha ! what's this, blocking up the way? Oh, yes,—brattice-cloth. Brattice, Miss Inquisitive,—brattice-cloth. Eh ? You—b-r-a-t-t-i-c-e,—there, now, did you get it? Oh ! that's it, eh? Laugh away, youngsters, for the laugh is certainly against me this time. I beg your pardon, young lady, for not waiting until I found out exactly what you were going to ask. Why, it is thick, heavy material, as you see, and is hung across openings throughout the mine, as it is here, to keep the fresh air current up in the breasts where the miners are. If the passages were all open, the fresh air sent down from above, instead of flowing along in the places where it is wanted, and where it must be sent, would go off into all parts of the mine. All through the mine you will meet with these brattice-cloths, in some places ; with wooden doors, in others ; while in yet other places you will find that former openings, which are never likely to be needed again, are closed up with solid masonry. To more fully insure the proper ventilation, or airing, of mines, the law requires that a man or boy be stationed at each important door, whose only business it shall be to open the doors, as necessity requires—for the passage for cars, etc.—and, particularly, to *close them again, properly*. When coal is to slide down one of these shutes, to the gangway, the brattice cloth is raised at the bottom just enough to let the coal pass under. We will go under it, too, but in the opposite direction, and proceed.

Here we are in the heading—the passage from which the rooms all open. As these breasts, along here, are all carried far up toward the chain pillar, we will not make the long climb necessary to get into the breasts where the work of mining is going on, but will see that work, after awhile, in some of the breasts just opened, farther out. The miners reach their working place by going up this passage, between the pillar and these timbers,—the *manway*, it is called. This timbering—posts and planks—you will find on both sides of the breast, and extending throughout its entire length. Sometimes the posts stand straight up, between floor and ceiling ; but, more commonly, they are placed slanting, like these, the upper end resting against the pillar, and the lower end on the floor, both ends held in holes made to receive them. The planks are then put on, making the manway, as you see, a kind of three-cornered tunnel. The miners have to drag all the timbers and other supplies, from the gangway cars, up through the shutes and manways to where they are needed—not very easy work, on such a very steep-pitching seam as is this. The timbers are put in as the work of mining goes forward, or upward, only a small space at the top being left open for the passage of the miners, and for other purposes. Aha! those are just the questions I've been looking for,—why is the timbering placed here? and where is the loose coal? If you could see through this plank partition, you would find the space inside full of loose coal; and that is just what these partitions are built for, up each side of the breast,—to hold the coal, as it is mined.

You see—but let me draw you a picture, which will help me in my explanation. Stand close together so I shall have enough light to draw the

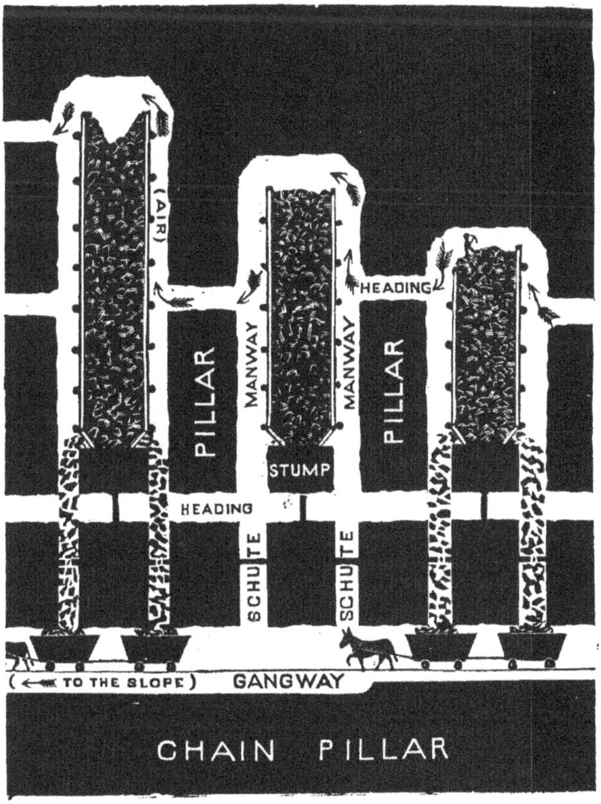

BREAST BEING WORKED.

picture, and you enough to see it. Here is the gangway, which we have just left; here are the shutes, through one of which we passed in coming from the gangway; here is the heading; and here are the breasts above it, with the pillars between, and the stump, here, beside us, where the timbering begins. Now, we will make the partitions up each side of the breasts. The round spots will represent these heavy posts, to which the planks are made fast. I can't show them, in this picture, as leaning against the pillar at the top, but we'll understand them to be thus placed, and also that the manways here at the side,—I'll mark them—which appear to be square openings, are really three-cornered openings, up under the slanting partitions. I have made the breasts of different lengths, you see, as corresponding with the difference in the times of their being opened up, as the gangway and heading were driven further. But we will imagine, however, that these rooms have not been opened up, as here shown, at all, but that the work is just to begin,—all being solid coal above the heading here. In opening a breast the miners first make narrow openings, on each side, up along the pillars,—exactly across the heading from the shutes, you see—leaving a section of coal—the stump, here—stand, for a short distance; then they open the breast clean across, just above the stump, keeping it that way, afterwards, throughout its whole length. In some mines, another plan is employed; no stumps are left, but the breasts are made the full width from the start, at the heading. All the coal taken out thus far, as the work proceeds, slides down the opening between the pillar and stump, across the heading, and on down the shute to the gangway. Once the breast is fairly opened out above the stump, the miners must begin to put in the timbers and line them with plank, so that they will hold all the loose coal taken out. Why? Don't you see why? The pitch of the seam is so steep that miners are not able to stand on the floor of the breast while they peg away at the hard wall of coal, in front of and above them; it would be like a man trying to stand on a very steep roof while he should work away, with heavy tools, at something fastened to the roof above him; he couldn't do it, of course. And so the miners, in steep-pitching seams like this one, (and some are even much steeper than this very steep one,) cannot work without something to stand upon, which will enable them to keep up to the wall of coal, before them. So, beginning on the stumps, they build these partitions, as they go ahead, allowing the loose coal to fill up the space, behind and beneath them, from which the coal has been taken. But they make more loose coal than they need under them, so they may pitch that not needed into the manways, above the end of the partitions, and it goes rattling and bouncing down the manway, (under the partition you know,) and on through the shute, to the gangway. Or, the loose coal filling the breast may be kept at the proper height to suit their work, by letting a part of it out, as circumstances require, through doorways made in the partition, at the bottom, just at the stump. This arrangement of the partitons thus built to hold the loose coal, or part of it, until the breast is completed, is called a *battery*.

I'll make the batteries, in our picture, full of loose coal—there!—to show you how the breasts look, thus filled. Up in this one, on the top of the loose coal, I'll make a miner at work, as we would find them if we were to climb up these steep manways to where they are now working. Down at the bottom we will open the door to let out enough coal to give him more room to work, at the top; here it goes, down the shutes to these cars which I have made standing in the gangway. This first breast, we will suppose, is completed, the coal all taken out up to the chain pillar. The breast now is full of the loose coal. We will open the doorways and let it all run out, so much at a time, down the shute and into the cars, to be taken to the slope, and up to the surface. These lines across the shutes and heading represent the brattice-cloths and doorways, which keep the air current on the right road. I will make another little picture, now, in which I will show the timbers leaning against the pillars,—there. Now, these three-cornered white spaces are the manways running up under the timbers, between the pillars and the loose coal in the breasts. You may keep these pictures, and show them to the other boys and girls, when you get home.

I may as well explain, now, how the air currents are kept up to where the miners are at work—and I can use my picture, in explanation, too. These narrow openings, running across the pillars, from breast to breast,

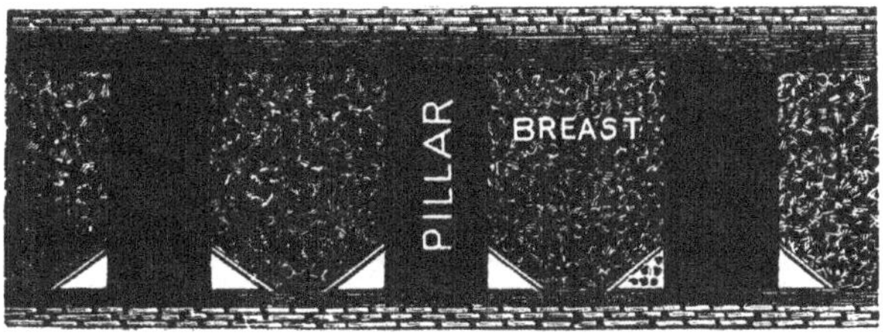

"PILLAR" BREAST.

are the *cross-headings*, pointed out in last night's picture. Cross-headings, from breast to breast, are frequently opened, you notice; but as each is opened, farther up, the one below it must be closed by doors or masonry. The fresh air is driven through the air-way, adjoining the gangway, to the farther end of the workings, where it is turned up into the breasts. These partitions, forming the battery, are made air-tight in planking them up, so that the air, as it is turned into a breast, must follow up along the manway, around above the ends of the partitions (which carries it across the breast where the miners are working,—just what is wanted,) down the other manway to the first cross-heading it finds, through this to the next breast, and then up the manway and across the breast, at the top, as before, and so on, untils it gets back to the slope or other opening through which it passes up to the surface again. These arrows will show its course through headings, along manways and across the breasts, at the top, where the work is going on. If an accident occurs, through the falling of the roof, for instance, by which the passage of the air through the proper course is checked, thereby cutting off the supply from the breasts beyond, it is, by properly arranging doors and brattice-cloths, made to pass down around the breast or breasts in which the accident may have occurred, and up into the next breast and then on, as usual.

Let us move on, now, along the heading, and see what is being done at yonder point, where the lights are. Ah, yes; here you can see what I have been showing in my picture—the running of the loose coal out of the battery, through the openings at the stump. A considerable loss is occasioned in the passage of the coal to the gangway by reason of the pounding and crushing and grinding of the pieces, which makes a lot of dust and very fine coal, for which but little use has as yet been discovered; so it is thrown upon the waste dump, those odd-looking black hills, which first caught Miss Inquisitive's eye. All the gob, too, (the slate, bony coal, etc., forming the partings in the coal seam,) is mixed with the coal, where the pitch is so great as to necessitate mining by this method. If there were enough gob in the seam to alone fill the battery it would be left there, the good coal being thrown off into the manway as

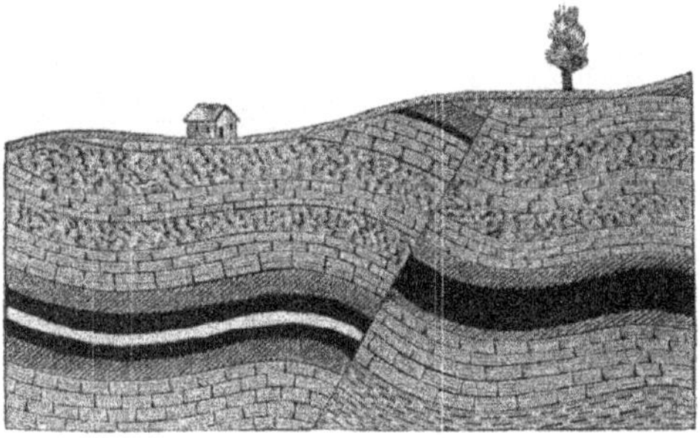

A "FAULT."

fast as broken up, and sliding down it to the shute—as already mentioned. In flat workings, buggy or wagon breasts, the gob, as picked and broken loose from the coal, can be thrown to one side, and be left there, as we have seen, the coal being sent out as fast as made ready. But here, gob and all must be sent out, as there is no way to separate them,—that work being left to the boys and men in the breaker.

Here we are, now, nearly to the end of the gangway, which is still being driven further ahead. Sometimes the straight-ahead order of things, in the way of working, is seriously interfered with, or even stopped, by the changed position or arrangement of the *strata*—as in a 'fault' or other displacement of the coal-bed. In the case of a fault, the miners unexpectedly find the coal-bed at an end, and in front of them a wall of some kind of rock, instead,—as I will show you by a picture. The *strata*, you see, have been broken squarely off, by some old-time earthquake, or other great movement, and the ends have slipped upon each other—one up or the other down, or both—so as to carry the *strata* out of their regular order of continuance, and to bring the ends of entirely different *strata* together. In such cases, the forward work must either be stopped entirely, or openings be made, through the rocks, until the coal-bed is again found, when work may again go forward.

Sometimes, too, a slate or other parting in a coal bed may increase from a few inches to several feet in thickness, dividing the bed into two more or less widely separated parts, or benches, each thick enough to be worked alone. In such a case, the owners of the mine would prefer to have the parting separate the two benches by a considerable distance, 40 or 50 feet, as then each bench could be mined separately; whereas, if they are not separated by more than a few feet, this cannot be done, and all the rock found between the two benches, must necessarily be taken out, too. Again, the coal may gradually grow thinner, from the top downward, and come to end, not by reason of a fault, but because the beds of vegetation which formed the coal, after once having been deposited, were swept away by strong currents or waves. It is thus sometimes missing, (other *strata* taking its place,) over a greater or less area. Then there are the 'squeezes,' which, occurring before the beds of vegetation became hardened into the solid coal, brought the rock *strata* above and below the coal close together, squeezing the coal out at these points, entirely or partially, and causing it to swell or bulge, at other points, to two or three times its usual thickness. These thick and thin places are often met with in mining. All these variations in the position and condition of the *strata*, affect the work of mining, as you can see. Sometimes, too, the coal at points in the seam will be found all crushed and ground up, and looking like so much black sand. This has been caused by some great movement of all the *strata*, the coal being at some points thus reduced to this fine condition by reason of the tremendous crushing and grinding force of the movement.

Let us go up into one of these breasts but recently opened up, and see what it looks like. We must go up the manway and get into the breast above the end of the partition. Give me your hand, Bess; now, up we go!—and here we are. It is in these places that the regular mining is done, day after day, until the limit of the breasts is reached, in each case; new ones being opened up as the old ones are completed. Drilling and blasting,—drilling and blasting,—with what picking is necessary to pull down the loosened coal, and to break off the pieces of slate, etc.,—it is thus the work goes on. In very many mines, or parts of mines, it is not necessary that blasting be constantly kept up throughout the whole length of the breast, in order to loosen the coal and take it from the beds; but the coal, once the breast is opened up, and the loosening fairly begun, continues to 'run,' as it is called, of itself, breaking loose from the seam, without blasting or other work, and continuing this, to a greater or less extent, throughout the length of the breast. Where the coal is very free-'running' precautions must be taken to prevent the 'running' process from extending into and destroying the pillars between the rooms, and the chain pillar, above,—which would probably lead to such a general falling in of the roof, or other movements of the overlying *strata*, as would destroy all the workings.

Let us make another picture, a side-view along the seam, to take away with us, with these other two, to show how mining is done on a steep pitch, like this. Here is the gangway,—this wide opening down here; and up here is the heading. Here we will have the shute, between the two, with the raised platform at the bottom (to carry the coal over the side of the mine car, you know) and with the timbering, looking like a fence, along the side—only I have made it too high, by half, I see, to be right. Now, up above the heading here, is the breast, which we will make full of loose coal. On top is a miner, pulling down the coal which he has loosened by a blast. Suppose, too, we represent the door at the bottom of the breast, or battery, as being open, with the coal running out and sliding down the shute into a car, which we will show standing in the gangway. There! we have our picture finished.

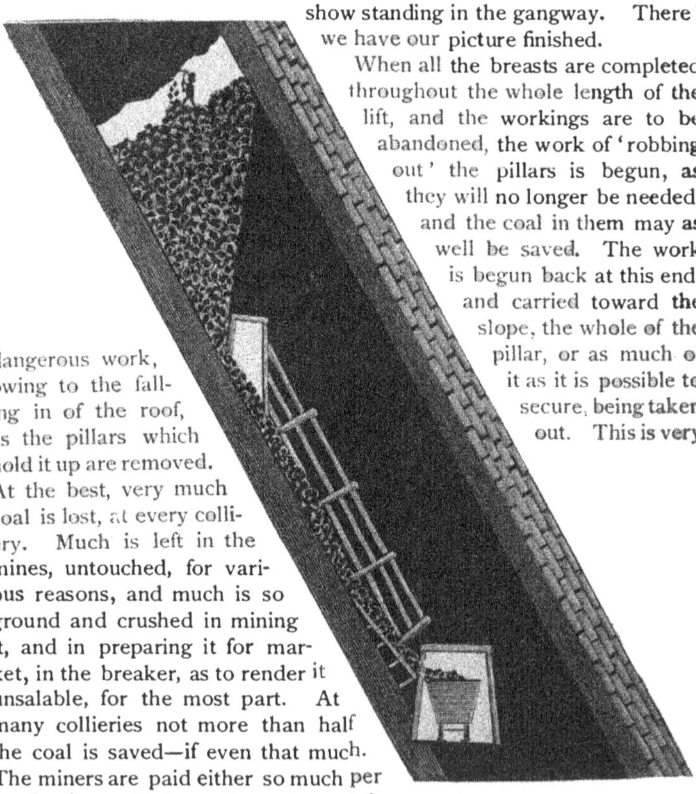

MINING ON A STEEP PITCH.

When all the breasts are completed throughout the whole length of the lift, and the workings are to be abandoned, the work of 'robbing out' the pillars is begun, as they will no longer be needed, and the coal in them may as well be saved. The work is begun back at this end, and carried toward the slope, the whole of the pillar, or as much of it as it is possible to secure, being taken out. This is very dangerous work, owing to the falling in of the roof, as the pillars which hold it up are removed. At the best, very much coal is lost, at every colliery. Much is left in the mines, untouched, for various reasons, and much is so ground and crushed in mining it, and in preparing it for market, in the breaker, as to render it unsalable, for the most part. At many collieries not more than half the coal is saved—if even that much. The miners are paid either so much per car for the coal taken out, or at a certain rate for the distance mined; either method of payment being employed as the circumstances require.

Let us, now, go down this shute to the gangway, and go back to the slope—a down-hill walk, all the way, I may say, for the benefit of you tired girls. Oh, no! the gangway is not perfectly level, but falls (though but very slightly—1 foot in, say, 150 feet, only) as we go toward the slope; this is to make it easier to pull the loaded cars to the slope, and also to cause the water to flow, readily, in the ditch, at the side. The water (and it is found in perhaps all mines,) is collected in what is called the 'sump,' at the foot of the slope, or shaft, as the case may be, to be pumped to the surface, by big steam pumps.

Well, here we are at the slope, again. Let us now go farther down—where the seam is nearly flat, and buggies are used, I am told, by our guide. Don't want to go down? Well, you must be pretty tired, as we have had a long tramp, winding in and out, as we have done,—and a hard tramp, too. So, if you think you understand, from last night's explanation, how mining is done in flat seams, as we would find below, we will not go down. We will sit here a little while, resting, before goin the surface again, and watch the trains coming and going, with the funny mule teams in front. We can watch, too, the loading and unloading of the cars at the slope, as the full ones are sent up, and the empty ones are brought down."

XIII.

MINE ACCIDENTS:—EXPLOSIONS AND FIRES.

"That man, over there, directing the moving of those cars, is the 'mine-boss,' who has general charge of the mining operations. He has his assistants, or other 'bosses,' having each his especial duties; as the 'stable-boss' and 'fire-boss.' The fire-boss? Why, his duty is to keep a careful watch throughout the mine, to prevent accidents which might otherwise occur, through the firing of gases which collect in parts of the mine. It is his duty to inspect every part of the mine, and to take prompt measures to clear the workings of large and dangerous accumulations of the 'fire-damp,' as the explosive gas is called, and to put out, or confine to a particular quarter, at least, any fire which may occur. Did you see that black board with the figures and white crosses on it, as you came in, this morning? That is the *fire-board*, as it is called. The fire-boss must examine every breast before the men come to work, to see if there is any fire-damp present. He makes notes of the search, and then prepares the fire-board, accordingly. He places on the fire-board the number of each breast, (they are all numbered, I should have told you before, and the miners are spoken of by their numbers, very often, instead of by their names,)—he places on the board the number of each breast in which he has found fire-damp, and places opposite the number one or more crosses indicating the quantity of fire-damp present. Each miner goes to the fire-board, when he comes to work, and there learns whether or not any fire-damp is in his breast, and if there is, how much. One cross means that there is a little present, and he must drive it out before going to work; two crosses mean that there is a considerable quantity present, and that extra precaution must be exercised; while three crosses mean that there is a dangerously large quantity present, and the miner must not enter his room, until given permission to do so by the fire-boss. Of course, if a miner's number is not on the fire-board, he understands that his breast is free from the dreaded fire-damp.

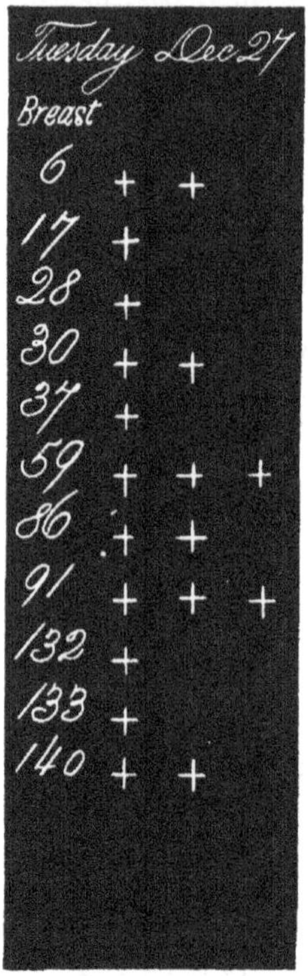

FIRE BOARD.

Frightful explosions sometimes occur, by which many lives are lost. The force of the explosion is very great, carrying everything before it, throwing the men flat upon the floor, and hurling them about, with mules, cars and timbers, also, as the terrific rush of air sweeps through the gangways and other passages. Frequently, too, the mine takes fire from the burning fire-damp, and a large part of the workings may be destroyed before the fire can be checked.

Fires often occur in piles of gob, and from the gob may reach the coal. The loose coal, in breasts, and elsewhere, is often destroyed in mine fires, but the solid seam of coal, itself, may not be much injured, as it is too compact to burn freely. A fire cannot make much headway where it has only the solid coal before it—no gob, loose coal or timbers. The smaller fires may be put out by the use of water, or, when in a pile of gob or loose coal, by removing all that is not already burning, out of the reach of the fire. When these methods do not avail to put out the fire, it may be confined to a certain quarter, and smothered, by closing up all openings leading to it, thus cutting off the air; for fire, you should know, cannot long continue without fresh supplies of air.

The fires following a great explosion are the most to be dreaded, and sometimes the whole colliery, including the shaft or slope, and even the buildings at the top, are destroyed,—though the shaft or slope, with buildings, may be thus destroyed, too, by fires starting in the stables or blacksmith shop. By the explosion of fire-damp another, and very poisonous gas, called the 'white-damp,' or 'after-damp,' is produced, and renders an attempt to enter the workings a very dangerous undertaking. 'Black-damp,' or 'choke-damp,' is another gas to be feared. When a fire in a mine is too large to be extinguished by the ordinary methods, and the whole mine is threatened, other methods are employed in the hope of checking it. The mine is 'sealed'—that is, all openings into it are closed up with masonry; and then steam is poured into the mine through pipes from the boilers. When all other plans fail, 'flooding' is resorted to; in other words, the mine is filled with water. Flooding is only resorted to when all other plans fail, as I have said, because it is a very costly operation of itself, while the damage to the mine by the water, is also very great. The flooding of mines, not counting the damage to the mine, has cost as much as $150,000! And then, too, after the fire may be put out, all the water poured into the mine—a vast quantity, of course—must be pumped out again; and by the time this is done, and all the damages to the mine repaired, many months have been lost and vast sums of money spent.

So, you see, the fire-boss has a great responsibility resting upon him, as, indeed, has every man working in the mine; a little carelessness may cause a very great loss. These little lamps of ours are the ones generally used in the mines, and you can see how easily the gas could be fired, by a miner carelessly going into a part of the mine containing it in considerable quantities. Old and abandoned workings are often full of the gas; and to avoid accidents, the openings into these places have been tightly closed. The proper circulation of the air-current through the mines, prevents, to a large extent, the collection of the gas in the workings in which mining is in progress. There are lamps made with a fine wire-gauze covering the flame, to prevent the gas from coming in contact with it. Such lamps are called 'safety-lamps,' but, I understand, they are not reliable safe-guards,—the miners, and the mine-owners preferring the ordinary lamps.

Other mine accidents occur, too. There are mine-car accidents of various kinds; powder explosions; accidental explosions in blasting, before ready; roof-falls; 'crushes'; and other lesser accidents of many kinds. A 'crush' is a serious affair, and is always guarded against, but sometimes not sufficiently. It occurs when there is not sufficient support to the rock *strata*, above, in the way of heavy chain pillars, and other pillars. It is not a simple falling of the roof where the support is not good, but it is a sliding out of its place of all the rock *strata* above the coal, crushing the pillars and everything else by the enormous weight and pressure, and ruining, beyond hope of repair, the whole mine, or a large part of it.

Some of the mine accidents are attended with unpleasant consequences to outsiders, too, occasionally. It is not unusual to find in the papers an account of the sinking of a considerable area of land, overlying a coal mine,—parts of mining towns often being included. The surface may sink several feet, in some instances, the people flying from their homes in fear that the houses may be thrown down. Usually these sinkings occur over old workings from which the pillars have been 'robbed.'

Taking it altogether, the life of a miner is not one to be envied. His work is hard, and the pay for it does not seem to be always satisfactory, if we are to judge by the numerous strikes occurring in the coal regions. He is surrounded by dangers—there are dangers on every hand. We should hold in thoughtful remembrance, as we sit at our ease by our comfortable fires, the hard-working, danger-daring men, whose labor makes such enjoyment possible to us.

Now, if you are quite rested, we will go up the slope, again, and pay a visit to the breaker and the rock and dirt dumps;—it is coming, at last, Miss Inquisitive."

XIV.

A COAL BREAKER.

"Back to daylight, again! and how good it seems. What a contrast to the gloomy place we left, a moment ago. I don't believe any of you would enjoy living in a coal mine; eh?

Well, as it is growing late, we must go over to the breaker, without delay, and after looking through it, we will go out upon the dumps, and then trudge home, again. We will follow this car, and see what becomes of the coal it contains.

Now, watch the car, as it reaches the end of the track. Tip! over she goes, almost upon end,—the front end down. Out goes the coal in a jiffy,—down upon this sloping iron grating, which lets the smaller pieces fall through, as it all slides along, while the large lumps go on down. We needn't try to follow this particular car-full any further, you see, as it has already been divided, and been mixed with other car-loads which were just ahead of it.

These immense buildings, called breakers, (for a reason which you will shortly learn,) are erected for the purpose of carrying on in them, properly, the work of preparing the coal for sale—for market, we usually say. As it comes from the mines, it is, as you have already seen, quite unfit for use, being full of slate, etc., and at some mines very wet and dirty looking; besides, pieces of all sizes are mixed together, many lumps being entirely too large for general use. In the breaker, all these matters are set straight, the slate, and other impurities are removed, the wet and dirty coal (where it is so), is washed, the big lumps are broken up—Yes, Miss Inquisitive, you've guessed it; that is why it is called a breaker.

A COAL BREAKER.

The big lumps are broken up, as I was saying, and the coal is finally so assorted as to bring all pieces of the same size together.

I do not know how high this breaker is, but some of them are as much as 115 feet in height. Few are less than 80 feet high. They cost, with the necessary machinery for properly carrying on the work of cleaning and separating the coal, a very large amount of money—more or less of course, in proportion to their size, and the amount of machinery in use. The smaller ones cost $25,000; larger ones, $50,000 to $80,000; while the immense structures at the largest collieries, cost as much as $100,000! Some of these large breakers can prepare 2,000 tons per day, though perhaps never actually doing so, as yet.

What a rattle and clatter is being made, inside, there; let's go in and see what is going on. Whew! the coal dust, it's nearly as dark in here, because of it, as it is down in the mine, even if the breaker is nearly all windows. Some breakers are, indeed, veritable 'glass houses,' but the people who live in them (during working hours) do not seem to regard the advice addressed to people occupying such houses, namely, that they 'should not throw stones;' indeed, hundreds of them, you will find, spend the whole day in doing nothing else but throwing stones—or, at least, pitching them. What do I mean? Just follow me, and I'll show you.

Here, now, are the 'slate-pickers,—the people who, as I said, throw stones all day long. Here, you see, is the coal—the larger pieces—as it comes from the mine cars. This long iron trough into which the coal slides from the iron bars which separated it, is called a shute. It carries the large pieces of coal to this platform, you see, where it is looked over and assorted by these 'platform men,' as they are called. They pick out the slate and rock, and throw them into this hole, which is the mouth of a shute carrying the slate, etc., to cars, to be taken to the rock-dump. As no slate is to be allowed to go into the coal, as prepared for sale, these men must examine every lump, and any slate found in any of them, must be taken out by the use of pick or sledge. Good, smooth lumps, of the proper size, are thrown into this opening—the 'lump shute'—to go to the railroad cars for shipment—this size being called 'lump.' Rough lumps go down this opening, as you notice, being carried to the 'crushers;' while all the small and fine bits go down here. So you see, now, these men, as are others in other parts of the building, are kept busy throwing stones, as it were, all day long. Now, let us go below the platform, to the crushers.

If the crushers were not in motion you would see that these iron or steel cylinders, or 'rolls,' have strong points—'teeth,' they are called—sticking out, all over them, and which break the lumps as they fall into the crusher. These large crushers or rolls are also called 'steamboat rolls,'—'steamboat' being the name of the size of coal next to 'lump,' in selling; it is, however, a size not made at all at many breakers. Where made, as at this one, the steamboat coal goes to the cars, through the shute, here, without further preparation. These rolls over here, are the 'prepared-coal rolls.' They are smaller than the steamboat rolls, and are used in preparing smaller sizes of coal commonly used. The smallest rolls are called the 'pony rolls' or 'monkey rolls.' How the lumps disappear before those strong teeth! reappearing, over here, in many pieces. What an appetite for coal these rolls seem to have, greedily eating up all that comes into their mouths and, like Oliver Twist, always asking for more! Let us next go and see how the pieces are separated, so as to get all the pieces of any one size away from the rest and into a heap by themselves,—which is done by 'screening.'

Here, now, are the 'screens,'—those great, long cylinders, covered with holes, and turning round and round. Oh! oh, they are hung lengthwise, this way, or, flat, as you call it, (turning on that iron shaft, which runs through their whole length,) because they could not do their work if standing on end. One end, you notice, is a little lower than the other, which causes the coal to slide slowly along, as the screens turn round and round. In this first screen, up here, the holes at the upper end are very small; those nearer the middle, a little longer; and those at the lower end, still larger. The next screen has three or four sizes of holes, too, all being larger than those in the upper one. All the screens are thus made. Only the dust and very fine particles go through these small holes or meshes here, you see, the rest sliding over them to the next size, where slightly larger bits fall through, and the rest again going on, as before, to the lower part of the screen, where the next size falls through, while all the larger pieces run out the open end and through that shute, into the next screen. Here the same operation is repeated, each size falling through as it finds holes large enough, until all has been separated. You see the different sizes dropping out,—each at its place, all along. All these screens separate the coal in the same way. The sizes here made, cover all those in general use, the smallest being 'buckwheat,' (not separated at some breakers, but going with the dust to the dump, as slack;) the next, 'pea;' then 'chestnut;' 'small stove;' 'stove,' and 'egg.' It is to supply the demand for these various sizes, smaller sizes that the lumps are broken up, as we have just seen, by the rolls. Here we meet again with the smaller pieces of coal which we lost sight of, you recollect, as it fell through the bars, when dumped out of the mine cars; all that smaller coal, and there is much of it, goes through these screens, to be separated.

But the pieces of slate must yet be taken out of the coal that has just been screened, as it has not yet passed through the hands of pickers,—as has the large coal, at the platform. The slate is removed by hand-picking or by running the coal through the 'jigging' machines,—over yonder. These machines have water inside, and as the coal goes in through the hoppers, the dirt, slate and other impurities nearly all go to the bottom, and are carried off, while the clean coal runs out into the shutes,—around here. Each size goes through a separate machine, of course. Some pieces of slate may slip through with the coal, but this man, here, removes all such intruders as they appear, and the clean coal runs down the shutes, into great bins, or 'pockets,' each size into its own pocket. If not thus 'jigged,' the coal, as it comes from the screens, is run through long shutes, called 'telegraph shutes,' lined with old men and boys, who pick out the slate as the whole slides along;—here they are, you see, some sitting astride the shutes. Sometimes, the picking is not done in the shutes, but on tables upon which the coal runs; in either case, the coal, when picked over, runs down to the proper pockets. Let us go down to the pockets.

Here they are,—great bins, into which the coal from above is constantly rattling, coming down through the shutes. Around on the other side of these pockets are the railroad tracks. They are lower than the bottom of the pockets; and the coal pours into the queer, hopper-bottomed railroad coal-cars from the pockets, over a broad shute,—or 'apron,' it is called, in this case. But let us go around and see the loading in progress. Here we are; now, you can see how it is done. This, of course, is where the work of shipping the coal is carried on. The breaker is always so located that tracks can be easily run into the lower side, from the nearest railroad. As fast as filled, you see, the cars are moved on a little, or run out, and others take their places. The weights are taken; a card, showing where and to whom the coal is to go, is tacked to the side of each car; a number of the cars are made up into a long train; and off they go,—some to owners of factories, and some to coal-dealers, in distant cities and towns.

The breaking of the coal, while necessary, in order to reduce it to the sizes most in demand, is a very wasteful operation, much of the coal being crushed into fine particles, which go to the slack dump. Let us leave this busy, noisy, dusty place, and go out to the dumps.

'Regular mountains,' you think; yet there are dumps, in other places, which are much larger than these, covering many acres, and being of great height. Indeed, it is a serious question, what to do with all the waste material from the mines, one-fourth to one-half of the material taken from the mines, being refuse, I have seen it stated. Formerly, all refuse—slate, dirt and the fine coal, or slack,—was thrown in one heap; but it is common practice now to keep the slack in a separate heap. This is now done because it is not unlikely that some ready method of using the slack may be discovered, so that it may all be sold, at a fair price, perhaps, instead of lying, unused, in these heaps. The sale of the slack, at a fair price, would add largely to the cash receipts of the coal companies, as immense quantities of slack are being heaped up at every large colliery. Some methods of using the slack, in connection with bituminous slack and tar, have been successfully employed, the mixture being pressed into rounded pieces of any size desired. This compound makes a very good fuel, but it has not yet come into general use. The distance from which the bituminous slack must be brought, adds largely to the cost of manufacturing this compound fuel; a method which would avoid this would be very acceptable to the coal-operators, you may be sure. Some forms of furnaces, for burning it in locomotives, are in successful use. It will be a long time, however, before the now waste slack of the anthracite region is all utilized.

What a fine view we get, from here, over the town! Quite a number of collieries are in view, too,—all at work, sending out the 'black diamonds,' to make cheerful thousands upon thousands of homes, and to keep fires roaring in blast furnaces and iron mills, and the wheels and spindles flying in factories. What a vast army of men, and women, and children, too, are dependent for their employment—their daily bread and home comforts, in effect—upon the hard, black substance coming from these mines! What a wise provision has been made for us, in the storing up, in the form of coal, of the sunshine of other ages!

The anthracite region is a busy part of our country. The mining industry gives employment to thousands upon thousands of men and boys, and has caused to spring up a large number of cities and towns. We have here seen but a small part of the region, but we cannot extend our visit to other parts of it, unfortunately. As we have now seen all the operations connected with mining in the anthracite region, and as our holiday time is limited, we cannot stay here, longer, or go to other parts of the region; but we will be off to the bituminous coal-region, in the morning.

Now, for a lively scramble down the side of this mountain of slack, and then homeward."

XV.

ON THE MONONGAHELA—TOW-BOATS AND COAL-BOATS.

"Isn't this a great way to visit coal mines—on a steamboat? Seems funny, to be sure; and yet, no better plan could be followed, to enable one to secure a good general idea of the Monongahela Valley coal mining and shipping operations. One must needs go to the Pennsylvania region to see anthracite mining operations, as extensively carried on; but we might have gone to other parts of Pennsylvania, or into Ohio, Indiana, Illinois, or any one of a number of other States, to see bituminous coal mines. I have brought you here, however, to the Monongahela Valley, because I wanted you to see what you can see nowhere else, so well, the method of shipping coal by river, instead of by rail, as it must be shipped from other fields. That you will be pleased with your visit here, I feel certain; the beautiful scenery along the rugged shores, to say nothing of the many strange sights of other kinds, upon and along the river, should certainly please you, very much.

While we are waiting for our boat to start, I may as well tell you something about the history of the Monongahela river and coal mining industry.

Do you see that black band along the hillside away up there near the top? That is what is known as the Pittsburgh coal-bed which I have before mentioned. It is from this bed that all the coal along this river, or at least, for a long distance up, is taken. The river is lined with coal mines, their black mouths being seen, in the hillsides, at short intervals, all the way up, for a hundred miles. They are all drift mines, too, the openings being made straight in from the face of the hills, on a level, or nearly so. Remember, we have left the tossed and tumbled-up anthracite country, and that here, on this side of the Allegheny mountains, the coal-beds and all the other *strata*, lie smoothly spread out, though dipping slightly, as you know, toward the Southwest, and being arranged in a series of several great waves, with wide troughs or basins, several miles across, lying between—the anticlinals and synclinals of which I told some of you, a year ago, when we were among the oil wells, now not far distant from us.

The fine quality of this coal was known away back a hundred and twenty-five years, or more, ago, when Pittsburgh wasn't Pittsburgh yet, but was only a lonely French fort and Indian trading post—Fort Du Quesne, in command of Colonel Boquet. Yes; quite a funny name, to us. We find frequent mentions made of the coal and coal mines, at that day. With the increasing settlement of this part of the country, by the white people, the use of the coal increased and mines multiplied, of course. With the opening up of steamboat navigation on the western rivers, in 1811, the increase in population and manufacturing was rapid, and, of course, increased the demand for coal. In 1817 the carrying of the coal in flat-boats was commenced. The coal would be mined during the winter, and sent down the hill and piled up along the shore, to await the breaking up of the ice and the coming of the freshets in the spring, when the coal would be loaded into flat-boats called 'French Creeks,' (because built on French Creek,) and sent down the river. They were simply allowed to float with the tide, and pretty much at its mercy, too, though the men in charge of each boat could guide its course by means of the great sweep or guiding-oar attached.

After this fashion the coal was floated to points down the Ohio river, for many years, the business steadily increasing all the while, as the towns and cities grew and multiplied, mining operations being gradually extended up the Monongahela. During the dry seasons, however, that stream was often reduced to a mere creek, interfering with the floating of the coal away from the mines, which could be accomplished only at certain times. To remedy this, a company was formed to build dams across the river, with locks through which the flat-boats could pass. Two locks were ready in 1841, and since then six or eight more have been added. They are from 10 to 15 miles apart, and each dam backs the water as far up as the next one, keeping it deep enough in each 'pool' for the passage of the boats. This has greatly helped the coal business ever since, you may be sure.

It was not until 1845, however, that the idea of 'towing' the coal-boats by steamboats built for the purpose, was put in practice. Like other progressive ideas, it was 'pooh-poohed! and laughed at, and those who advocated it were called fools. But a successful trip was made to Cincinnati, and from that time to this, the towing of coal has been a steadily increasing business, until now there are great numbers of tow-boats, and thousands of the box-shaped vessels in which the coal is carried—*barge*, *coal-boat* and *flat* being the names given to the different kinds of these coal-carrying boats.

The coal tow-boats, as you have seen, are not like the fussy little 'tugs' which some of you have seen at Chicago, or New York, or elsewhere, pulling the great vessels after them, by means of a cable; but are very large and powerfully built vessels, with tall smoke-stacks, and with a great paddle-wheel behind. Barges are large, flat-bottomed and straight-sided vessels, 130 feet long by 25 feet wide, and 5 or 6 feet deep, and holding 13,000 bushels of coal—oh, my! no; coal is not 'measured out' by the bushel,—though it is sold by the bushel; being weighed, instead of measured, however, to find the number of bushels in a wagon-load, or car-load, or boat-load, as the case may be. In the anthracite region, all sales are in tons, you know; but here, while there are sales by the ton, of course, nearly all sales are by the bushel. The law says that 76 pounds of coal shall make a bushel; and so, all we have to do is to weigh our coal, to get at the number of bushels. In loading railroad cars or coal-boats, each mine carload is weighed before being dropped into the car or boat; and so, by taking the weight of all the loads emptied into it, we can tell how many bushels are in a car or boat, you see. Families buy this coal by the bushel, as weighed, the price here, for best coal, bring five to six cents per bushel. But to proceed: The flats are much like barges, but smaller, holding 4,000 bushels. Barges and flats are carefully and substantially built, and may be used for nine or ten years—unless sooner sunk or mashed to pieces by some accident. They are always brought back to the mines, after the coal is unloaded where sold. Down the river,—perhaps at New Orleans, over 2,000 miles away, by river,—boats are the largest, holding 24,000 bushels. They are intended for just the one trip, being more cheaply built, than the others, and are sold with the coal, down the river. Then there are the fuel-flats, (holding about 7,000 bushels,) which carry the coal to be used by the tow boat, itself, while making a trip.

Just now, as you see, there are quite a large number of tow-boats, lying, 'tied up,' along the shore, here in the harbor, waiting for a rise in

WAITING FOR A RISE. WANTED—WATER.

the river;* for, while the dams and locks permit of constant passage up and down the Monongahela, they, of course, cannot help the condition of things in the Ohio, which is now at a low stage. Besides the boats lying here, there are many more scattered at various points down the Ohio, unable to come up, with their 'empties,' until there is more water. Then, too, there are others, lying at the mines up along the Monongahela. You see, there are two classes of tow-boats. The smaller ones are for use on the Monongahela, only, taking the loaded barges, etc., from the mines to the harbor, at the head of the Ohio, at Pittsburgh, and taking the 'empties' up to the mines again, from the harbor, as brought back from 'below,' as it is termed,—that is, down the Ohio and Mississippi. The larger boats take the coal from the harbor to the cities and towns down the Ohio and Mississippi; some going, as I have already intimated, as far as New Orleans. Much of the coal is brought down from the mines, during the times of low water in the Ohio, and is made fast—'tied up'—to the shore, or to the piers of bridges, by great chains and cables, to await the coming freshet. There it lies, you see, acres and acres and acres of it, along the shore, in the barges and other craft. Besides these, large numbers of loaded barges, etc., are lying at the different mines, up the river, to be hurriedly brought down, when the rise begins, and turned over to the larger boats, for the trip 'below.'

The greatest feat on record in boating was that performed by one of the largest tow-boats,—the W. W. O'Neil,—in May, 1886. The tow of this boat consisted of forty-one barges and other boats, loaded with coal to the amount of 796,820 bushels, and two other barges, loaded with 400 tons each of hay, etc.,—forty-three pieces in all! The tow was 302 feet wide and 640 feet long,—covering an area in the water of nearly four and one-half acres! the coal thus being towed along representing the quantity (pillars, etc., being left) from nearly eight acres of the Pittsburgh bed. To give you an idea of the quantity of coal in the tow, let us have Ned figure it out for us, supposing it were to be hauled by railroad. Each car, we will say, will hold 300 bushels of coal, and each locomotive can haul twenty-five cars; now, how many trains will be required to haul it? Ah! let's see,—that's right,—2,656 cars! or 106 trains of twenty-five cars each! Quite a load of coal for one boat, you see.

The system of towing coal is most novel, and is peculiar to itself, in the principles involved and the methods employed. As a recent writer

---

* As a matter of fact, up to the date of this writing, December 27th, there has been no general "run" of coal down the Ohio since early in the summer—a condition of things without precedent in the history of coal-boating, and having been brought about by the drought which has been so general throughout the Union. Even the dams on the Monongahela have failed to afford a sufficient depth of water, and an additional height of 18 inches was given them, by means of a temporary arrangement of planks, in order to run the coal from the various works down to the harbor, at Pittsburgh—where 13,000,000 bushels of coal, in barges, etc., are waiting the long-deferred rise; all barges, flats, etc., on hand being full. A "coal-boat stage" of water is from 10 to 12 feet above low-water mark, and a number of "runs," of from 6,000,000 to 10,000,000 bushels, each, are usually made each year. No system of transportation, it may be stated, can compare in cheapness with the towing of coal, as here practiced. The cargoes handled by the tow-boats—comparatively small vessels—are greater than any ever carried by that famous ocean steamship, the "Great Eastern."

says, 'No other transport service in the world is based upon similar principles of construction or navigation; and he further adds, that it is a striking fact that the ablest general in the British army, intended, in an approaching expedition to the upper waters of the Nile, to employ two stern-wheel boats, built upon the ideas underlying the construction of these tow-boats. We shall, no doubt, have an opportunity to see a tow-boat with its tow of barges, etc., 'under way,' before we go elsewhere; indeed, we may be fortunate enough to see many of them.

Ah! our boat is backing off from the wharf, I notice."

XVI.

UP THE RIVER: COAL WORKS—DAMS AND LOCKS.

"Off we go! at last. And what a cloud of the blackest smoke, we leave behind us, as the coal is thrown into the furnace under the great boilers! No such a display of smoke as this, in the anthracite region; was there? And after all, while the smoke and soot of the bituminous coal soils and makes dingy everything white, one likes to watch the great billows of it pouring from the smoke stacks of steamboats, and locomotives and factories, and the smaller columns of it rising from the chimneys of the houses. How the paddle-wheels swish and swash as they go round and round in the water, shoving our boat forward. We have two of them, one on each side of our boat, while the tow-boats have but one,—behind; our boat is a 'side-wheeler,' while all the tow-boats are 'stern-wheelers,' you notice. Some passenger boats, however, are stern-wheel boats, instead of side-wheel boats, like this one.

Yonder comes another boat, down the river, meeting us. Run together? Oh, no Bess, you needn't be afraid of that; they will pass each other, safely. No, indeed, no danger of that either; each knows which side to take. How? Well keep your ears open now; Listen—our boat is whistling. She gave one blast, you notice. Now the other boat is whistling; one blast, too, so—no, there's another, making two blasts. Our boat is whistling again; two blasts, this time. Now, see, both boats are turning out of their course, a little, and will soon pass. Don't understand, Ned? Well, the pilots were talking to each other by these blasts of the whistles. Our pilot told the other one when he blew the first blast, which side he wanted to take, in passing; or, we may say, he was asking the other pilot if we might take that side. You see, when boats meet, the boat going down the river always has the 'right of way;'

DAM AND LOCK.

that is, the pilot of the boat going down has the choice of sides, in passing. The boat going up whistles one or two blasts, as the case may be, to tell the other which side it wants to take. If it suits the other boat, it answers by giving the same number of blasts, and the arrangement is settled; if not, it gives the opposite number, and as it has the right of way, the boat going up has to answer the upper, giving the same number, and pass to that side, accordingly. The law requires this exchange of signals, so that collisions cannot occur. The boats must answer each other until both blow the same number of blasts, which shows that they understand each other. There she goes by us—a tow-boat, but without a tow. What waves she piles up in her wake, as her big wheel churns the water! In passing, where one boat has a tow of coal with the barges loaded down until the top is but little above the water, it is necessary for the other boat to go very slowly, lest the waves from her wheel should sink some of the other boat's barges.

COAL PIT MOUTH—SCENE ON THE MONONGAHELA.

Yonder is a colliery,—or *coal-works*, they call it, here,—on the western shore. Away up yonder, near the top of the hill, you can see the mouth of the mine—'pit-mouth,' we say, here, 'coal-pit' or 'pit,' instead of 'coal-mine' being most generally used. Some cars are just being hauled out, you see, from the 'pit,' following a track laid out along the hillside to the *check-house*—that dingy-looking building perched against the hillside. You see there is a railway—an 'incline'—from the check-house down to that building on the shore. In the check-house is winding machinery,—drums, etc.,—similar to that which we saw at the anthracite collieries. When the loaded cars from the mine reach the head of the incline, at the check-house, the wire cable is attached to them, and

PIT, CHECK-HOUSE, INCLINE AND TIPPLE.

they go flying down the incline, some empty cars going up the other track at the same time—there go some, now. That house, built out over the river, is the *tipple.* The loaded cars run into the tipple, where they are tipped up, the end opens, and out goes the coal. It falls upon screen bars like those we saw at the breaker, the smaller pieces going through, and the lumps sliding down into that large swinging iron pan, we will call it, which you can see, from which it is dumped—into the river? Oh, no; but into that big barge, which you see lying under the outer end of the tipple. The smaller pieces fall upon another set of bars, or screen, and are again divided, the larger pieces, called 'nut' coal, going into another barge or flat, and the slack into yet another,—you see the two, further under the tipple. While being filled, the barges or other boats are held in place by great cables, made fast to posts in the shore. As the vessel is filled at one point it is allowed to float down a few feet, being then again made fast while the coal is emptied into it. When full, it is floated to a point below the tipple, and made fast alongside of other loaded ones, another one being floated into its place at the tipple.

The breaker? Oh, there are no breakers in the bituminous region. This coal does not have to be broken for use, as the lumps thrown upon the fires burst into a blaze, at once, and in a few minutes, when slightly warmed, can be split and broken, if it is necessary or desirable, by a blow of the poker. For the open fireplaces--grates—everywhere seen here and wherever bituminous coal is used, lumps as large as your head, or even larger, are preferred to the smaller coal. And what a beautiful and cheerful fire they do make, as we saw, last night,—the yellow blaze roaring up the chimney, and the gas-jets pouring out of the cracks of the lumps, some as long plumes of soft white smoke, like little volcanoes, blazing like miniature gas wells; others, and—but I'm running clean away from the subject in hand. There are no breakers, as I said, and all the separation of coal that is practiced is into slack, nut and lump; the nut coal is about as large as unshucked walnuts, while lump coal includes every size from this up to the great big pieces which you couldn't lift. Sometimes there is no separation, at all, all going together. Mining here, with drift mines, and with no breakers, is not so expensive as in the anthracite regions.

But look ahead, yonder. Aha! what is it, indeed? Yes, it does look like a waterfall Ned, sure enough; and it certainly is one; but not a natural one. It is one of the dams I told you about, stretching across the river and blocking our way, it seems. Oh, we'll get up, Miss Inquisitive, —somehow. We are heading toward the shore now, you notice, where those walls are; that is the lock, and it is there that we will find a way to get up over the dam. Now, see how it is done. The lock-keeper has heard our whistle and is getting ready for us. See those two great doors at the lower end of the walls, opening back against them; now they are back and our boat can run in—for that is where it is heading. There is another pair of gates, you see, at the upper end, kept closed, now, to keep the water from the dam from pouring in between the walls. Now, carefully—in we go! the top of our boat only level with the tops of the walls! Now they are closing the big doors, again, behind us; we are shut in, sure enough. But look yonder; they are opening the upper doors, and—oh, no, no—we won't be drowned, Bess, you goosey; have you forgotten that our boat floats on the top of the water, always? Come and see how we get out of this trap. The doors are opening slowly, you see, and the water is, of course, coming in. Higher? Ha, ha! how astonished you look! Of course we are rising; the water from above is pouring in, and filling the space between the walls,—for it can't get out, below, now, you see—and we go up as it rises. There now, the lock is full, up to the level of the dam; all we have to do now is to run on up the river—which we are already doing! For boats going the other way the plan is reversed. The upper doors are opened, the lower ones being kept shut, until the boat runs in off the dam—between the walls,—the tops of the walls, that is the remainder being hidden, as when we ran out

RIVER COAL TIPPLE.

just now. Then the upper doors are closed, and the lower ones opened. The water between the walls runs out, the boat getting lower as the water falls, until it is where we were when we first ran in. It is then at the level of the water below the dam, and runs out and goes on its way! Without these dams, which hold the water and keep it deep enough for the boats, we could not be boat-riding now, as the river, on account of the dry weather, would be without them a mere creek.

A RAILROAD "TIPPLE."

Coal-works, you notice, are numerous, on both sides of the river. They are all alike, in general, too, having the incline down the steep, rocky, hillside from the pit, the tipple at the river, and the barges, loaded and empty, lying along the shore, with one or more tow-boats with them. See what a slack-dump at that one ; it runs, you see, from the pit, almost at the top of the hill, down to the railroad track, at its foot.

Suppose we go ashore here now, as the boat is going to land to take on some passengers ; we can stay here, to-night, and to-morrow visit one of the mines.

Off with you,—quick ! but don't fall off the plank, into the river.

'Rain, rain, rain !' This will please the coal-shippers, I know. It has been raining since about ten o'clock, last night, and if it continues at this rate until to-morrow morning, we'll see busy times on the river, sure. Now, put on those waterproof coats and cloaks, and we'll visit yonder mine,—if it is a bad day.

Whew ! that was a climb,—sure enough ! Steep as a ladder, I've heard people say ; and though I don't know just how steep that is, I suppose this must be about what is meant. But let us go into the mine ; it isn't raining in there.

No shaft or slope to ride down, here ; on the contrary, we just walk in, as upon a floor, its a little rougher, of course, but nearly as level. And so we would find it, throughout the mine, were we to go all over it. By the way, I may say that in some of the older mines here, along the river, the coal has all been taken out, long ago, and the coal which one sees coming from the pit-mouth now, comes from mines back in the other hills, sometimes quite a distance away,— the old mine opening now serving as a tunnel to get the coal from the new mine to the river.

All through this region, and not alone upon the river, there are coal mines,—nearly all being drift mines, in hillsides ; but there are some shafts at points where the coal has dipped below the hills. Coal-works line the railroads, and a great shipping business is carried on. The railroad tipples are like these river tipples, the coal—lump, nut and slack—being loaded into cars just as it is here loaded into barges and flats.

The miners here are not paid as in the anthracite region ; but all their coal is weighed, and they are paid accordingly, the rate now being three cents per bushel,—that is, 76 pounds. All the coal is weighed in the tipple, before going into the barges or railroad cars. Each miner is known by his number—the number of his room, that is—and is provided with a lot of short pieces of wood bearing his number. One of these pieces he puts in each car of coal he sends out of the mine. At the tipple, this numbered piece of wood is taken out, and the owner is credited with the weight of coal in the car. When pay day comes, at the end of the week or month, he is paid for the amount mined, as shown by the tipple records. In order to protect the miners, there are two weighmen in the tipple,—one the company's clerk, and the other, called a 'check-weighman'—representing the miners.

The miners are of many nationalities—Americans, Englishmen, Welshmen, Scotchmen, Irishmen, Frenchmen, Italians and others. There are more foreigners than Americans, by far. There are many thousands of miners in these coal fields.

Now, we are getting back to the workings. Not much work is being done, now, however, (and, indeed, none at all, at some mines,) as all the barges on hand are loaded, and empty ones cannot be brought up from down the Ohio, until the river rises. Here are breasts—or rooms, they are more often called here—on both sides of the gangway, or entry, as it is commonly called. The pit-cars are run right into the rooms to be loaded, you see, and there are no shutes nor batteries as in the anthracite mines. We are now at the end of the workings in this direction, I find — though they are still being pushed forward. Here is a room just being opened, the miner working his way with a pick instead of a drill, for blasting, as in the other coal-region. He is 'bearing in,' now, along the side of that block of coal, having already cut back under, at the bottom. It is only little by little, indeed, that he cuts his way into the black mass before him; but, once he has finished the bearing in, he can, by wedging at the top, bring down a great mass of coal. As, however, we have already seen in the anthracite mines practically the same operations we would see here, we will not stay longer, but will go to the surface, again.

Still it pours! Once down this slippery path and at home, we will sit by the fire for the remainder of the day. To-morrow will bring us some sights."

XVII.

A COAL-BOAT RISE.

"Oho! look at the river, this morning! That flood of yellow water means—all coal out for 'below.' We will catch the first boat, and be off down the river.

All aboard, again! Be quick, Ned. Off we go, this time down the river. Now keep your eyes open, and see what the coal-men are about. Away up the river, yonder, are two or three boats in sight, each with a tow of barges and flats. Others are following them, as those great clouds of smoke beyond the bend testify. Look over here, at this landing. These two boats are busy making up their tows from the barges lying here,—the boats backing and pushing and choo-chooing, and the men shouting and running about like mad. Everybody is in a hurry, you see; and boat rises don't come every day. We are slowing up now, to pass this boat with its tow. And here is another, just in front, too. Notice

TOW BOATS AND COAL FLEET "TIED UP" READY FOR A RISE.

how the tow is made up. On either side of the tow-boat is a flat—the fuel flats—while the main part of the tow is in front of the boat, one set of barges in front of the other, making it two barge lengths in depth, and spreading out two or three times the boat's width, too. Those cables crossing from one to the other, hold the barges together side by side and end to end.

Here is another boat, with its tow, just getting under way,—and here is another to be passed; we can travel faster than they, you see. See the smoke beyond that bend; there must be several boats close together, just ahead. Look! there must be half a dozen, or more! And they do not seem to be moving! Oh, I have it; we are at the lock, and the boats are waiting their turn to go through, 'First come, first served' is the rule at the lock. It takes some time to get a boat and tow through, as there is room for but so many barges, etc., at a time; and then, besides the time taken to pass them through, other time is lost in separating them—breaking up the tow—and in getting them properly rearranged after the passage. Before double locks were made at the lower dams,—all the boats and tows having then to pass through one, instead of being divided between two, as now,—there were some lively scenes at this lock. The river would be fairly filled with boats, all wanting to get through as soon as possible, lest the water should begin to fall again before the tows could be started down the Ohio. Some had to wait a day and night before their turn would come, and quarrels sometimes occurred, caused by some captain trying to steal for his boat another one's turn.

Now we are through, and away, with plenty of company, in the way of tow-boats, around us. Here we are in the harbor again. What a different scene from that which we left, two days ago. The river is alive with boats, pulling and pushing, running in and out and up and down—and whistling! Well, you never heard the like; did you? And smoke! you never saw the like, either. Coal-fleet is in motion; the smaller boats bringing down the coal, as we have seen, and the big fellows, here, getting it into shape for a start when the water rises yet a little higher, as it will do when the full effect of the far-up streams is felt. See that large lot of coal lying over there, with the three big tow-boats lying alongside. They are getting up steam in the boats, you see, to get their tows ready—and big ones they'll be, too, to include all those barges and flats. We will go to dinner, as soon as we land, and come back this afternoon in time to see them go out.

Just in time! they are beginning to go. See yonder fellow, bravely tugging away to get the tow past that pier. Ah! too late! crash it goes! and is crushed like an egg-shell by reason of the tremendous momentum or force, of the huge field of coal. Down it goes, to the bottom! See the men cutting the cable to let it loose; now, the rest swing under the bridge, and away they go! The company may charge the lost barge to 'profit and loss'; and maybe there'll be some others to be entered on the books thus, before they all reach their ports. Mishaps often occur, not

DOWN THE OHIO—TOW-BOAT AND FLEET ON THE WAY.

only one, but sometimes many, barges are lost, or delay is occasioned—the striking of bridge piers, collision with other tows, 'swamping' of barges, running upon sand-bars and sticking fast, etc. Great skill and equal coolness must be possessed by the pilot. He must know every foot of the river, so as to be able to follow the often narrow and tortuous channel—the deepest part of the river, where the boats must run—as it shifts from shore to shore, and must carefully calculate his steering as he swings past obstructions or between the piers of bridges. Ice is a source of danger to the coal men, at the time of a general break-up. When suddenly breaking away before the pressure of the dammed-up water, after having been gorged for miles, perhaps, it sweeps everything before it, even destroying or greatly damaging the strong dams. A few years ago, the ice in a few minutes crushed thirty steam-boats and many barges and other coal-boats!

Away they go, one after the other, as their tows are made up. The dwellers in the far-away cities will welcome them, as it means lower prices for coal than they have been paying for some time. Let us see another off, and then go; we can't see all off, for some may not get away before midnight. Here she comes, the largest tow-boat in the service, too. And what a tow she has! A great island of coal! I do not know how much coal she has this time, but one of her former tows was the largest ever taken out.

I read, the other day, a fine description of the handling of one of these immense tows of coal, and I cannot do better than give it to you. The writer says: 'The "driving," (for such it almost seems to be, in its handling by the deft pilot who, with sinewy arms, whirls and rewhirls the wheel that guides the boat and this mass of coal,) is a task to which only those brought up to the trade are competent. Skill, judgment, nerve are all called into play as this ponderous bulk, borne along on a river at flood height, running at a current of eight to ten miles an hour, sweeps onward. Through narrow channels, round sharp bends, between the stone piers of bridges, where a mis-turn of the wheel, a failure of judgment, a miscalculation of distance means disaster and wreck, the pilot guides the tow; now backing, now flanking, now pushing, now floating, watchful and cool, the pilot does his work. There is probably no such boatmanship shown anywhere else in the world as is displayed by the Pittsburgh coal tow-boat pilot. Watching one of these ponderous tows surging down the river with the tow-boat of but perhaps 90 to 100 feet in length and 20 to 25 feet in width at its rear, turning it round bends, flanking it past points, backing and checking it in narrow channels, one cannot but think of the old joke of the tail wagging the dog, and here it does it, and does it well. It is a wonderful exhibit of skillful navigation, the thus handling by the nervy grip of one man on a wheel, a bulk of 30,000 tons, moving at a speed of from 12 to 15 miles an hour down such a tortuous stream as the Ohio, and with perhaps not five feet to spare of channel width or two feet of water depth.'

Now, let us go home. To-morrow we will visit the coke-region."

XVIII.

IN THE COKE REGION.

"Here we are in the great coke region—one of the most wonderful regions of our country, in many respects; one certainly not having its like anywhere. It is in this part of the coke field, known as the Connellsville region, which is only about fifty miles long by three miles wide, that the business of coke-making is carried on most extensively; but a very large coke business is carried on at various points in the extension of this field to the north and east, along and upon the Allegheny mountains and far up toward the northern and central part of the State—this region being, you know, in the southwestern part.

See those great clouds of smoke, rising skyward, beyond that hill-top; we will make for that point, as that smoke is a sure indication that we shall find there what I want you to see—coke ovens. Oh, no, young lady, I hadn't forgotten those long-standing questions,—and I may as well answer them now, as we walk along.

Coke is made from coal—bituminous coal—by burning it in *ovens*, built for the purpose, of stone and fire-brick. These ovens are usually of the shape of the old-fashioned bee-hive (which you may have seen in pictures, if no where else), and are, therefore, called 'bee-hive' ovens. Never saw one? Well, if you were to cut that orange into halves, and lay the pieces down, on the flat part, you would have the shape of the ovens, pretty nearly. That is, the inside of them is of that shape; outside, you only see stone walls, with a flat top, scores of ovens being built together in a 'range.' There are other forms of ovens in use, but the bee-hive is the most common one.

While coke can be made, and is made, at many points outside the coke region (any bituminous coal, with the exception of the 'block' coal, being used), yet the coal of this region is peculiarly adapted to the making of coke. The coal of this region is very much softer here than elsewhere in the same bed; so soft indeed, as to unfit it, before being coked, for many manufacturing purposes. It can, therefore, be very easily mined, it being a lighter task for the miner to pick it loose from the bed, than to shovel it into the pit cars, afterwards. Not much like anthracite, you see. When the coal is brought—ah! now the ovens are in sight; let us go closer, climbing yonder bank, where we can have a good view, and be in nobody's way.

Now, we have the whole cokery spread out before us, and an odd scene it is, too—and a busy one. Yonder are other ranges of ovens, too,

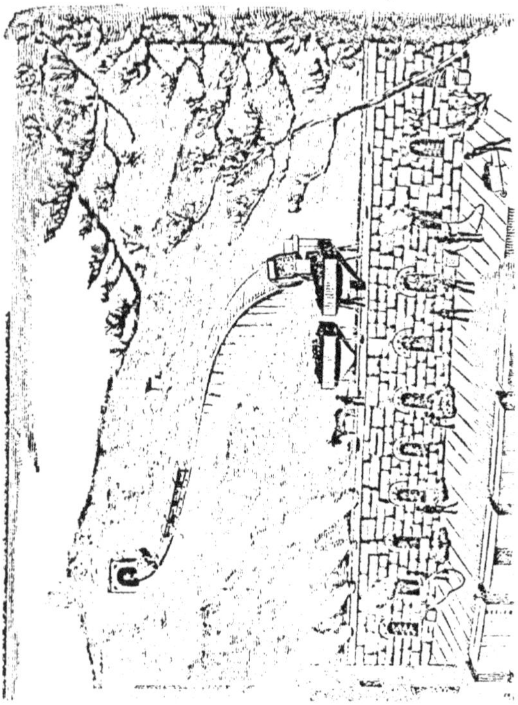

—off to the right. We might go for miles along the railroads (which, unfortunately, we do not have time to do), and everywhere find these long ranges of ovens. They are always built along the railroads, or where switches can be run into them, to provide for the easy and rapid loading of the coke, as fast as made.

Yonder is the mine from which the coal for these ovens is taken ;—there come some cars, now, down the incline to the ovens. The men yonder, empty them all, as that one is being emptied, now, into those big hopper-like iron cars—called 'larries'—which run on the track laid on the top of the ovens, being drawn by horses or mules, and at some places, by small locomotives. Each of those little doors in front opens into a separate oven, as is the case with those openings on top, from many of which the smoke and flames are pouring. The ovens are 11 or 12 feet in diameter and 5 or 6 feet high, and are entirely separated from each other, by thick walls. Yonder, now, is one being filled, the coal running out of the larry, through a door just opened at the bottom, and directly into the oven through the hole at the top. About 100 bushels—which is 7,600 pounds, you can see—is put into each oven for one 'charge', being spread evenly on the floor. It is made two feet deep when it is to burn for two days—48 hours—and 2½ feet deep when it is to burn three days—72 hours—as when in over Sunday. One-day, or 24-hours, coke is sometimes made, but the average time of burning is 48 hours, this giving about the best coke.

Whew! the smoke! how it is pouring out the top. You see, the first coal put into an oven, as in this case, takes fire from the heat left by the charge just removed, and which you can see lying on the platform, below the door. As fast as an oven is emptied it is filled, again; and thus it is kept going all the time. The front opening, you notice, has been almost entirely closed up with bricks coated with wet clay, to shut off the air; as the coking goes on, the opening is entirely closed, and later, the one on top, also.

Yonder is one just being opened, the coking having been completed. What a white-hot furnace it is, inside! Now watch that man as he throws water upon the coke in the oven, by means of that hose and pipe—for it is cooled before being taken out. What a cloud of steam it makes! Now, the man with the long rake is getting ready for his part of the work—and out comes the coke, tumbling upon the platform. In a little while the larry will be brought up, and that oven filled again. Everywhere along the range you can see these different operations going on—for one oven is opened, when ready, regardless of the others. Some are being filled, others emptied, you see, and other men are busy loading the coke into cars to be sent north, south, east and west.

That tall building? That's the *crusher*,—not so large as the anthracite breakers, but serving much the same purpose. Those great pieces of coke are what are wanted at blast furnaces, and at other such places, and it is for their use that the great bulk of the coke is made; but of late there is a demand for coke for household uses, and it is to properly supply this demand that the crushers are built, to break the coke into smaller pieces.

The especial value of coke lies in its great firmness. The coal from which it is made, and all bituminous coal, excepting the block coal, melts down when burning, and cakes; while coke will bear, without crushing, an enormous weight of iron ore, iron, etc., in blast furnaces and iron mills. In coking, the oxygen and hydrogen escape from the coal, leaving the coke nearly pure carbon, like anthracite,—nearly 90 parts in 100 being carbon. It is of a beautiful silvery color, as you can see, and—look at this little piece—full of very small openings. It has a sharp clink or ring,—as you notice when the men move it about.

There are many thousands of coke ovens, built this way, in ranges, and a night ride through the coke region affords a novel sight. The industry, as you here see it carried on, gives employment to thousands of men,—millions of tons of coke being turned out yearly. Coke was made here seventy-five years ago, but it is only within the last twenty or twenty-five years that its especial value has become known and the business has grown to such an extent. Coke is made at many mines outside of what is known as the coke region, the slack, which would otherwise be wasted, being used in its manufacture. Coke is made in some foreign countries, also.

Coking, as now carried on, is a wasteful process. An enormous volume of gas goes up into the atmosphere every day, in the clouds of smoke everywhere rising. So great is this volume of gas, that there has been serious talk of forming great companies to lay immense pipes to carry it to the eastern cities—Philadelphia, New York and others. It is estimated, too, that $20,000,000 is lost yearly in this district, in tar, ammonia and sulphur given off by the coal, and which might be saved by the use of some other process of coking.

Talking of gas suggests an idea. We will not stay here until to-morrow, but will take a train back to town, and close our sight-seeing by visiting the gas-works to-night."

XIX.

SOME FIGURES.

"How much coal mined this year? I'm sorry, Ned, that I can not tell you; the year is just out, you know, and the figures covering the amount of its business, in any line, may not be published for many weeks or months, as it seems to require a very long time to get them together. I can, however, give you the figures for 1886, which, of course, will be reasonably near those for this year—though perhaps a million or more of tons less.

The following figures for 1886, for our *whole* country, are given by our Government officers:

	Tons.	Value at mines.
Anthracite (Pennsylvania)........................	39,035,446	$73,706,957
Bituminous (all fields)........................	76,119,120	78,481,056
Totals........................	115,154,566	$152,188,013

More than one-third of all the coal mined, and worth nearly half the total value, came from the anthracite region, you notice. And, would you believe it? the first coal shipped from that region—to Philadelphia, in 1803—was pronounced worthless, as they had not then learned how to properly burn it!

By the way, the first coal-mining in our country was at Richmond, Virginia, coal being sent to Philadelphia, New York and Boston, in ships, as early as 1790. Anthracite was first used in 1768, by blacksmiths. Regular mining of anthracite was begun in 1820.

We need not be afraid of running out of coal soon. It is estimated that the anthracite field can supply us for 200 years with 100,000,000 tons per year! while the Pittsburgh bituminous bed is estimated to contain, in its 3,000 square miles, 20,000,000,000!—enough for 2000 years!"

"Although the electric light is already so largely taking the place of gas for lighting our streets, stores and houses, as you see, we cannot yet get along without the gas—and, of course, then we, must have gas-works. As I told you yesterday, on our way back from the coke region, gas that we burn to make light is made from coal—bituminous coal. In natural gas regions, it is true, the natural gas is burned for light, to a large extent, as well as for heat; but very largely, even in such regions, and entirely so, in others, the gas used for light—illuminating gas—is made from coal by burning it—as you shall see, up this street now, a few steps. Here we are; we will pass first into the retort-room, at the left.

How smoky and dingy! and what a queer looking place generally. Those odd-shaped iron things, sticking out of the brick wall and with pipes running up from them, Miss Inquisitive, are the ends of the *retorts*—all of which are now closed. The retorts? They are the places in which the coal is burned, to make the gas—as you will see later. These square doors at the bottom are furnace doors—fire being kept in the furnaces, below the retorts, to heat up the coal when the retorts are freshly filled. The—but let us watch that man yonder, who is about to open one of the retorts. The door, you see, lifts entirely off, being held in place by latches and that screw-arrangement. What a beautiful, white glowing mass is lying in the retort; it is coke, such as we saw made in the ovens, to-day. These retorts, you see, are, in effect, little coke ovens. Now he is pulling out the coke with his long iron hook;—hot work, you may be sure. Now we can see the shape and size of the retort; it has a flat bottom and rounding, or oval, sides and top. It is, I should think, about a foot high and runs back about eight or ten feet into the brick-work. Here he comes with an iron wagon full of coal; we will watch him fill the retort again. Whew! how quickly the coal bursts into a blaze as he shovels it in!—and look at the smoke pouring out the door! The retorts hold about 200 pounds of coal each, which will yield about

IN THE GAS-WORKS.

1,000 cubic feet of gas,—1,000 cubic feet being such a quantity as would fill a room or space ten feet wide, ten feet long and ten feet high; that is, it is a space ten feet in every direction. Now he has filled the retort and is about to put on the door—using another door, however, which has been prepared for the purpose by having clay spread over. The clay?—oh, to make the retort air-tight, which it must be.

By the action of the heat, as the coal is burned in the retorts, it is separated into the different substances of which it is composed. The solid part remains in the retort as coke,—just like what we saw made, to-day; for these closed retorts act like the coke-ovens. The other substances—the gas, tar, ammonia, etc., in more or less liquid or vaporous form, leave the retorts and pass up these large pipes, here in front, and into that big, sooty cylinder, upon the top of the brick-work, and which is all the time about half full of tar. From this cylinder the gas, tar and liquids containing ammonia, etc., pass out through that pipe you see running from the farther end of the cylinder through the wall—and which we will follow. Here, you see, it comes down into this little room, and goes on through the floor. There is a hole or well under the floor, called the *tar-well,* and the tar and much of the ammoniacal liquors are left in this well, while the gas, being very light, rises to the top of the well and then passes out through a pipe to the *condensers*—in here; there they are,—those big pipes doubled up and down in a long row, against the wall. The gas must pass through all these pipes, around the bend at top and bottom, and as it does so, it becomes cooled and some of the liquids which have still remained with it, as vapor, are condensed—like steam, the vapor of water, when it touches a cold surface—and run down into the box at the bottom of the pipes, to be drawn off. The gas itself, now more pure, passes on through the whole length of the condensers, and then by a pipe into the *purifiers*—these big square chests—where the remaining impurities are removed. Here is one open; it has, you see, a great lot of slatted shelves, smeared with lime, while lime covers the bottom also. Into one of these purifiers—the covered ones—I mean, though—the gas pours, as it comes from the condensing pipes, and winds in and out among these shelves before it passes out again, through a pipe, into the next one. The lime on the shelves takes from the gas the sulphur and some other impurities, and it is thus left fit for use in our houses and stores and on the streets.

But it doesn't go straight to the places where it is to be burnt, however; it is first all collected in the big iron tank or gas-holder—incorrectly called a *gasometer,* meaning a gas measurer, which the tank is not. But on the way to the gas-holder, as we'll call it, the gas does pass through a measurer—the *meter.* The meter, you see, is a sheet-iron box. It is partly filled with water in which is placed two cylinders divided into a number of long, three-cornered spaces by partitions running lengthwise through it. You will have to imagine this, as we cannot see it through the box. As the gas comes into the meter, through the pipe, it enters the spaces in the cylinder, and as it fills them turns the cylinder round and round. Each space in the cylinder holds so many feet of gas, together holding ten feet. Every time the cylinder goes round, ten feet of gas has gone through; this the meter shows by the pointer on this dial; and every tenth time it is turned (or hundred feet measured, that is) the pointer on this next dial is turned to show the fact; and so on, for thousands, ten thousands and hundred thousands of feet. So, you see all the gas made is here measured. As it is carried around by the cylinder, it passes out through another pipe and into the big gas-holders,—out here.

They are, you see, big, round, covered iron tanks, like the oil tanks, —only, instead of standing on the ground, each is floating in the water, in a big, deep, round, brick-walled hole, Did I say floating? Yes, ma'am, floating; you see, the gas pouring into them, which it does through a pipe coming up from the bottom, and standing above the water—holds them up, and as more and more pours in, lifts them up until the top reaches the catches on these strong iron pillars, standing beside each one. No, they are not so high, now;—you are right. But, you see, the burning of thousands of lights on the streets and in the houses has drawn out much of the gas, and so the tanks have been lowered; by morning they will be much lower, as the gas, by that time, will have been nearly all used out. By the pressure of these great iron tanks, or gasometers, the gas in them is forced out through all the streets in great iron pipes, or mains, and from these into the houses and stores, through small pipes, and is ready, at the striking of a match and the turning of the stop-cock, to blaze away, to our great convenience and comfort. Between the big street pipes and the burners in the houses—usually in the cellars—is placed a meter, like the big one we just saw, only smaller, which measures and records the number of cubic feet of gas burned in each house, and which the people in the house must pay for, at so much for each 1,000 cubic feet.

Now, let us go home, passing through the retort-room again. Another retort is being filled, you see; and so the work goes on—a steady stream of gas flowing from the retorts to the condensing pipes and purifiers, and on through the meter into the big gasometers, and then, as needed, out to houses and stores. This man is shoveling some of the coke into the furnaces under the sets of retorts; what is not so used is sold. A considerable quantity of ammonia—hartshorn—is also produced in gas-making, and is sold; as is, too, a smaller quantity of sulphur. There is not so much waste at gas-works as at coke-works, you see. Even the black, sticky, bad-smelling tar has its uses, and several of them; among them, one of which none of you would ever suspect it of being guilty, I'm sure. It is of no use to have you guess; so I will tell you. From this tar there is made the beautiful and brilliant aniline colors! the discovery of which has been of untold value to the world. Why, some of the tar in this very barrel, here, may furnish the substance which shall color the royal robes of some queen or princess,—or the gay ribbons of Miss Ruffleton, here! Astonishing, indeed. But come, we must be off.

Well, here we are at our lodging place again. We cannot stop to talk longer, now, as we must start for home early in the morning;—I'll answer your questions on the way, Miss Inquisitive.

Good night, then—all of you."

CONCLUDING CHAPTER.

A MAN After piloting the "Boys and Girls of America" through localities where coal is mined and prepared for market, and explaining everything that pertains thereto (within the prescribed limits of a useful Christmas story), would feel as if he had failed to perform his entire duty, if he neglected to call their attention to a few additional important facts. He would state before bidding them a final adieu for the year 1888, that they are indebted to what is popularly known as

"THE GREAT ROCK ISLAND ROUTE,"

or its General Manager, for the pleasure they have derived from the perusal of this and previous Christmas and New Year Annuals; an enjoyment (he trusts) to be realized with the advent of every coming year. This Souvenir (like those which have preceded it) is a compliment intended not only for the youth of our Country, but for their parents, aunts, uncles, older brothers, sisters and grown up relatives—all in fact who are kindly disposed to the **ROCK ISLAND** system, who have traveled over any of its main lines, branches or extensions, or contemplate doing so at any future time. **A MAN** would remind such friends and would-be patrons, that on no other railway in the world can such incomparable, elegant, convenient, luxurious and beautiful Day Coaches, Dining Cars, Pullman Palace Sleepers and Reclining Chair Cars be found; that at all terminal points including Council Bluffs, St. Joseph, Atchison, Leavenworth, Kansas City, Minneapolis and St. Paul, transfers are made in splendid Union Depots; that the **ROCK ISLAND** offers a choice of routes to California, the Pacific Coast and intermediate towns and cities west, northwest and southwest from Chicago, and in corresponding opposite directions; that it traverses the finest cultivated portions of the Middle West, including the great states of Illinois, Iowa, Minnesota, Dakota and Missouri, and has opened up new avenues of travel and traffic to all parts of southern Nebraska, interior Kansas and points beyond; that its Fast Express Trains move with uniform speed and regularity on a smooth, steel track, crossing mighty rivers over bridges of stone and iron; that its facilities and accomodations are unsurpassed; that its employes are polite, attentive and trustworthy, and that ladies traveling alone or even children, are as safe from harm (and always as comfortable) as in their own parlors at home, while under their protecting care. Hoping you will not forget these suggestive facts to which he briefly invites your considerate attention, all that now remains for **A MAN** to do, is to tell you, (if you desire to be more thoroughly posted), that you can obtain tickets, maps, folders, copies of the Western Trail, or whatever information you want in regard to rates, routes or connections by applying in person or by letter to

E. ST. JOHN,
General Manager.

CHICAGO, ILL.

E. A. HOLBROOK,
Gen'l Ticket & Pass. Agt.

—OR—

G. L. RHODES, City Passenger Agent, 104 Clark St., Chicago.
S. S. STEVENS, General Agent, 1305 Farnam St., Omaha, Neb.
MATT CLAIR, Traveling Agent, Omaha, Neb.
A. B. FARNSWORTH, General Eastern Passenger Agent, 257 Broadway, New York City.
W. J. LEAHY, Traveling Agent New York State 257 Broadway, New York City.
E. W. THOMPSON, New England Passenger Agent, 296 Washington St., Boston, Mass.
JOHN T. DOUGLASS, Traveling Passenger Ag't. 296 Washington St., Boston, Mass
I. L. LOOMIS, Passenger Agent, Middle District, 111 South Ninth St., Philadelphia, Pa.
JOS. A. SHEPPARD, Southeast Passenger Ag't. Chicago, Ill.
H. S. PHILPS, Traveling Passenger Agent for Quebec and the Maritime Provinces, 154 St. James Street, Montreal, P. Q.
D. L. CAVEN, Traveling Passenger Agent for the Province of Ontario, 35 Yonge Street, Toronto, Ontario.
C. H. HOLDRIDGE, N. E. Passenger Agent, cor. Larned and Griswold Sts., Detroit, Mich.
M. P. WASHBURN, Passenger Agent for the Northwest, Chicago, Ill.
CLINTON JONES, General Agent, 36 Montgomery St., San Francisco, Cal.
F. W. THOMPSON, Traveling Agent, 36 Montgomery St., San Francisco, Cal.
J. F. FUGAZI, Italian Passenger Agent, 5 Montgomery Ave, San Francisco, Cal.
CHAS. KENNEDY, General Agent, Rock Island and Albert Lea Routes, 3 Washington St., Portland, Oregon.
A. H. MOFFET, General South-western Passenger Agent, 532 Main St., and 1042 Union Ave., Kansas City, Mo.
W. H. FIRTH, Traveling Passenger Agent for the Southwest, 532 Main St., Kansas City, Mo,
JOHN T. SMITH, Trav. Pass. Ag't, 532 Main St., Kansas City, Mo., & 9 Windsor Bl'ck, Denver
C. C. STOWELL, Traveling Agent, Chicago, Ill.
WM. RICKEY, Passenger Ag't, Davenport, Iowa.
W. I. CHEEVER, Passenger Agent, Central District, Peoria, Ill.

Spotted something that isn't quite right?

Let us know and we'll fix it for future reprints. And as a thank you for your diligence, we'll send you a free Sothis book of your choice.

publisher@sothis.press

This page intentionally left blank

Made in the USA
Monee, IL
08 November 2026

68106067R00059